Sounding Conflict

Sounding Conflict

From Resistance to Reconciliation

Fiona Magowan, Julie M. Norman, Ariana
Phillips-Hutton, Stefanie Lehner, and Pedro Rebelo

BLOOMSBURY ACADEMIC
NEW YORK • LONDON • OXFORD • NEW DELHI • SYDNEY

BLOOMSBURY ACADEMIC
Bloomsbury Publishing Inc
1385 Broadway, New York, NY 10018, USA
50 Bedford Square, London, WC1B 3DP, UK
29 Earlsfort Terrace, Dublin 2, Ireland

BLOOMSBURY, BLOOMSBURY ACADEMIC and the Diana logo are trademarks of Bloomsbury Publishing Plc

First published in the United States of America 2023
Paperback edition published 2024

Copyright © Fiona Magowan, Julie M. Norman, Ariana Phillips-Hutton, Stefanie Lehner, and Pedro Rebelo, 2023

For legal purposes the Acknowledgements on p. vi constitute an extension of this copyright page.

Cover design: Louise Dugdale
Cover image: Sounding Conflict: A Performance in Five Acts / Pedro Rebelo, with Matilde Meireles in collaboration with Tinderbox Theatre Company © Pedro Rebelo

All rights reserved. No part of this publication may be reproduced or transmitted in any form or by any means, electronic or mechanical, including photocopying, recording, or any information storage or retrieval system, without prior permission in writing from the publishers.

Bloomsbury Publishing Inc does not have any control over, or responsibility for, any third-party websites referred to or in this book. All internet addresses given in this book were correct at the time of going to press. The author and publisher regret any inconvenience caused if addresses have changed or sites have ceased to exist, but can accept no responsibility for any such changes.

Whilst every effort has been made to locate copyright holders the publishers would be grateful to hear from any person(s) not here acknowledged.

Library of Congress Cataloging-in-Publication Data
Names: Magowan, Fiona, author. | Norman, Julie M, author. | Phillips-Hutton, Ariana, author. | Lehner, Stefanie, 1976– author. | Rebelo, Pedro, 1972- author.
Title: Sounding conflict : from resistance to reconciliation / Fiona Magowan, Julie M Norman, Ariana Phillips-Hutton, Stefanie Lehner, and Pedro Rebelo.
Description: New York, NY : Bloomsbury Academic, 2022. |
Includes bibliographical references and index. | Summary: "Examines the creative processes and contested politics of sound, music, and storytelling for conflict transformation"– Provided by publisher.
Identifiers: LCCN 2022043223 (print) | LCCN 2022043224 (ebook) |
ISBN 9781501383021 (hardback) | ISBN 9781501383052 (paperback) |
ISBN 9781501383038 (epub) | ISBN 9781501383045 (pdf) |
ISBN 9781501383069 (ebook other)
Subjects: LCSH: Music–Social aspects. | Music and conflict management. | Sound–Social aspects. | Storytelling–Social aspects. | Peace-building.
Classification: LCC ML3916 .S667 2022 (print) | LCC ML3916 (ebook) |
DDC 306.4/842–dc23/eng/20220909
LC record available at https://lccn.loc.gov/2022043223
LC ebook record available at https://lccn.loc.gov/2022043224

ISBN: HB: 978-1-5013-8302-1
PB: 978-1-5013-8305-2
ePDF: 978-1-5013-8304-5
eBook: 978-1-5013-8303-8

Typeset by Newgen KnowledgeWorks Pvt. Ltd., Chennai, India

To find out more about our authors and books visit www.bloomsbury.com and sign up for our newsletters.

Contents

Acknowledgements		vi
List of Figures		viii
Foreword		ix
Introduction: Sound ambiguities		1
Fiona Magowan, Julie M. Norman, Ariana Phillips-Hutton, Stefanie Lehner and Pedro Rebelo		
1	Sound methodologies in conflict transformation and peacebuilding	29
	Fiona Magowan, Julie M. Norman, Ariana Phillips-Hutton, Stefanie Lehner and Pedro Rebelo	
2	Resistance: Performing the frontline	47
	Julie M. Norman	
3	Resilience in creative practice in a post-conflict context: Musicians Without Borders	71
	Fiona Magowan	
4	Remediating relationships: Collaborative storytelling and conflict	97
	Ariana Phillips-Hutton	
5	From noises of conflict to dissonant sounds of reconciliation in the Northern Irish theatre	121
	Stefanie Lehner	
6	Working through creative practice: Socially engaged arts interventions	151
	Pedro Rebelo	
Conclusion		173
Fiona Magowan, Julie M. Norman, Ariana Phillips-Hutton, Stefanie Lehner and Pedro Rebelo		
References		189
List of Contributors		205
List of Partners		209
Index		215

Acknowledgements

This book is the result of the five-year Sounding Conflict research project funded by the AHRC/Partnership for Conflict, Crime and Security Research under Grant AH/POO5381/1 (see http://www.soundingconflict.org). There are many people who we would like to thank and without whom this research would not have been possible. Firstly, we are most grateful to Dr Olivier Urbain who served as an international expert adviser across the life of the project and whose perspicuous insights and wise guidance have been invaluable. The project has been ably supported throughout behind the scenes by Christina Captieux whose wide-ranging expertise has meant a smooth delivery of all our various outcomes. We would also like to thank Dr Jim Donaghey for his creative and social science contributions to the research, including interviewing, producing Storymaps and developing a creative montage. Finally, we are especially grateful to Professor Hastings Donnan for his ongoing support and expert guidance of the project and for hosting it at the Senator George J. Mitchell Institute, Queen's University Belfast.

In addition to our project team, we would like to thank all of our partners and the individuals who participated in the research and shared their stories and experiences. Firstly, we would especially like to thank Farah Wardani and all the members of the Laban theatre group for their collaboration, critical reflections and ongoing creativity. We are also especially grateful to Christina Foerch and the members of Fighters for Peace (FFP) for making introductions, welcoming us to events and dialogue sessions and sharing their insights on difficult subjects.

We would like to thank Musicians Without Borders (MWB) for their generous collegiality in supporting the research. We are indebted to Laura Hassler, director of MWB, who facilitated our collaborative research, providing expert advice and support. In addition, the director's assistant, Meagan Hughes, very kindly supplied us with various materials and hosted our three visits to the Music Bridge programme. We would also like to thank the senior facilitators and trainees of the MWB Music Bridge programme without whom research for Chapter 3 would not have been possible. They cheerfully and expertly guided us through the programme and warmly welcomed us into the MWB community. Our thanks go to all the participants who took the time to think deeply and critically with us, answering surveys, engaging in interviews, exploring creative processes and offering pertinent insights into their music making. We would also like to thank Eibhlin Ní Dhochartaigh and Peter O'Doherty for their assistance and for allowing us to undertake the research at the Cultúrlann Uí Chanáin in Derry/Londonderry.

We wish to express special gratitude to the Kabosh, Tinderbox and TheatreofplucK theatre companies and their artistic directors, producers and sound engineers, as well as the Rainbow Project, in particular Paula McFetridge, Patrick J. O'Reilly, Jen Shephard, Katie Richardson, Niall Rea, Isaac Gibson, and Dean Lee. Thank you for your generous support and invaluable insights in facilitating the research for this chapter. Many thanks to Laurence McKeown, Patrick J O'Reilly and Shannon Yee for making scripts available and for allowing us to quote from them. We are also very much indebted to our colleague David Grant for his vital help, advice and vision in organizing and facilitating the Image theatre workshops – and many thanks to all the participants of these workshops as well as all the audience members who actively and enthusiastically participated, provided important feedback, patiently filled out surveys and thereby offered important insights and reflections. We would also like to thank David Crooks for his kind assistance and for allowing us to use the facilities in the The Dunanney Centre in Rathcoole. Finally, special thanks to Matthew Logue for invaluable help with photo and film editing.

We would like to thank Mary Kouyoumdjian, Milad Yousufi, Yuki Numata Resnick, Yamilette García, Julia Cordani and all the students and families of Buffalo String Works for their generous gifts of time, energy, stories and, of course, music. Without you and all those who support you, this research would not have been possible.

We wish to thank Museu da Maré in Rio de Janeiro, Cláudia Rose Ribeiro da Silva, António Carlos Pinto Vieira and the participants in the Som da Maré project (2014).

Figures

0.1	left, Josefina de Vasconcellos's 'Reconciliation' (Stormont Castle, Belfast); right, Maurice Harron's 'Hands across the Divide' (Craigavon Bridge, Derry).	16
5.1	The perception of the RUC constable. Image-Theatre Workshop with members of the Rathcoole Community on 18 June 2018.	135
5.2	The Garda's perspective. Image-Theatre Workshop with members of the Rathcoole Community on 18 June 2018.	136
5.3	Actor Rhodri Lewis as Ubu delivering his victory speech, inviting audience participation, in *Ubu the King* (Tinderbox, 2019).	137
5.4	Actor Noel Herron, playing our guide, directing our attention to a woman across the road, played by Holly Hannaway in TheatreofplucK's *So I Can Breathe This Air* (2018).	146
5.5	Actor Martin McDowell 'targeting' actor-participant Richard Bailie, with our guide Noel Herron coming in the way, in TheatreofplucK's *So I Can Breathe This Air* (2018).	147
6.1	Sounding Conflict: A Performance in Five Acts: Still from Act I	168
6.2	Sounding Conflict: A Performance in Five Acts: Still from Act II	169
6.3	Sounding Conflict: A Performance in Five Acts: Still from Act III	169
6.4	Sounding Conflict: A Performance in Five Acts: Still from Act IV	170
6.5	Sounding Conflict: A Performance in Five Acts: Still from Act V	170
6.6	Sounding Conflict: A Performance in Five Acts: Live performance in Museu da Maré, June 2022	172

Foreword

Olivier Urbain

It was such a life-affirming, convivial and meaningful experience to be part of this five-year-long Sounding Conflict research project. This major contribution to peacebuilding praxis through sound-based creative arts is in many ways a celebration of socially engaged practice grounded in sonic trust. When we carefully listen to each other's sounds and voices, create spaces where our sonic vibrations are heard, and improve our communication, connection and relationships, that is when trust has a chance to emerge between us.

In the four aspects of peacebuilding carefully examined in this work, trust may be placed in a central position. It seems to be a crucial ingredient in each of these activities:

- Resistance: for community organizing and cohesion
- Resilience: to enhance our collective capacity and flexibility to manage disturbance
- Reconciliation: to be able to look at the past and towards a better future together
- Remediation: to find collective remedies through changing the format of artworks

We find the word 'trust' (as a substantive, a verb and a few times as trustworthiness and trusting) about fifty times in this volume, mostly in Chapter 3, which focuses on trusting relations in musicking, and two or three times in most of the other chapters (none in Chapter 2 and once in Chapter 6).

Trust starts with listening. Elise Boulding, one of the founders of peace research, famously declared that listening is the beginning of peace. When we listen and others are heard, we create the possibility for us to be heard, and for a deeper connection to be established, which can eventually lead to building trust. Most of us have experienced the joy of listening to each other's voices, and each other's singing and musicking. One major contribution of this book is to broaden the sonic scope to include all sounds, even noise, in this interconnected world of sonic vibrations. We hear this sonic festival expressed in storytelling, in theatre (drama, Forum Theatre or Playback Theatre), in collective musicking, in avant-garde musical compositions and in installation art, among other events.

At first I was sceptical as to the capacity of non-musical sounds and noises to contribute anything to listening, connection, trust and peacebuilding; with a few remarkable examples in this volume, I was quickly convinced of their effect on an audience, allowing for new ways to embrace discomfort, surprise and interruption: a doorbell ringing, a bomb explosion, a rifle shot, the belliphonic dropping of kitchen equipment on the floor, the drone of helicopter rotors and so on. High-pitched radio tuning can invite us to listen very closely, and the fading between two soundtracks can throw us in the liminal space between reality and fantasy. A guided performative audio-walk, an audio-journey through a familiar city can transform our understanding and perception, giving us a fresh perspective on the same streets, buildings and parks.

All this can enhance our capacity to listen more carefully, to doubt our entrenched certainties, to have deeper dialogues and, with perseverance and care, lead to more trusting relationships. It might be worth writing a sequel to Ian Middleton's 'Trust' keyword for music in peacebuilding, expanding the realm of sonic experience to non-musical sounds; with this volume, we now know what to listen to – to all sonic experiences, unconstrained by any musical or cultural filtering or prejudice.

The mixed methodology in this book involves listening and sharing through participant observation, narrative interviews, surveys, group discussions, practice-based reflective analysis and more. The case studies are mostly based in Lebanon, Palestine and Northern Ireland, but Brazil is very present too, offering the experience of the Som da Maré in Rio de Janeiro (2014) and inviting the voices of refugees from multiple countries through the compositions of Mary Kouyoumdjian: Afghanistan, Burma, Venezuela, Morocco, Malaysia, Eritrea and Mexico, all converging on Buffalo, USA.

A welcome aspect of sonic trust is that it can develop between academics from different disciplines who are willing to listen to each other and find common ground while accepting differences and collaboration as dissonant harmony. The authors of this book come from such diverse backgrounds as music studies (ethnomusicology, composition and performance, musicology and music philosophy), anthropology, political science, literature, drama and English. Our multidisciplinary, informative and emancipatory exchanges during our meetings in Belfast are fondly remembered as laughter, humorous statements, and good cheer, and even though the pandemic prevented us to be together for more than two years, I can still remember everyone's engaged and vibrant voice.

With so much sonic material available, so much goodwill from countless people and so much understanding of the primacy of listening, sharing and trust, should we be more optimistic about the prospects for a more connected, blossoming and peaceful humanity and biosphere? Not really, because this volume also contains sobering insights regarding the potential limitations of this type of research. In the introduction, we hear that relational and participatory arts activities are not always conducive to an inclusive, liberating and democratic experience, and need to be planned carefully. In the conclusion, we are reminded that the vertical power structures in place are rarely moved by all this horizontal sonic trust building.

Just as researchers have abandoned a naïve optimism regarding the automatic potential of music for peacebuilding (music is completely ambivalent, and it all depends on what we do with it), this work expands this scepticism to all sonic vibrations. But scepticism leads to grounded and careful explorations, and that is a sound approach to peacebuilding endeavors, as was shown in this volume. It is well worth a listen.

Introduction: Sound ambiguities

Fiona Magowan, Julie M. Norman, Ariana Phillips-Hutton,
Stefanie Lehner and Pedro Rebelo

Sound is fundamental to our experience of everyday life – a foundation which accrues deep significance in protracted conflict and post-conflict settings. Sound shapes people's perceptions of and responses to environmental and spatial risks, as well as intensifying experiences of (in)security among those living in areas that suffer the effects of violence. In this introduction, we advance a rationale for understanding sound as a crucial component of the potential for the sound-based creative arts to transform conflict. Although this may seem circular (if we hold that creative practices can contribute to conflict transformation, then sound must be important to that potential in those practices which work in and through sound), a closer look reveals that this is a comparatively neglected aspect within discussions of conflict transformation. In academic analyses of those creative practices which are closely linked to sound (e.g. music, storytelling), specifically sonic qualities have often been framed as secondary to verbal and gestural aspects. In artistic practices which are not immediately conceived as sonically driven (e.g. literature, theatre), sound has occasionally appeared as a subject of comment but only rarely as the centre of attention. This reflects an unspoken hierarchy within academic writing across the humanities and social sciences in which what is most significant is that which is most susceptible to being relayed through textual and visual means. It also reflects the inherent elusiveness of sound; in its ephemerality, sound is forever slipping through our grasp, undermining our ability to pin it into place for analysis. Yet even in the realm of verbal communication, *what* we say is always linked to *how* we say – and, moreover, how what we say resounds with our listeners.

Sound, in its immaterial materiality, is ever at the centre of communication, of relationship and of the potential for both conflict and its transformation. For example, the visceral impacts of sonic vibration have been shown to create connection and solidarity by deepening empathy and generating what neurologists call 'limbic resonance' (Lederach 2011: xi). At the same time, sound is critical for those who have experienced violence to be able to respond verbally and non-verbally to the interpersonal and sociopolitical effects of conflict. Thus, this book is about the ways in which the sonic participatory arts might cause us to hear something different and also to hear differently (following Brandon LaBelle [2021]) within the context of transforming conflict.

This volume explores modes of listening to, engaging with and articulating sonic dynamics. In doing so, we tease out the intertwined and often opaque nexus between sounds and their effects in theatrical and musical contexts. We consider the remit of sound to include non-verbal environmental sounds as well as humanly created sounds, such as musical instruments, the sounds of speech and, by extension, storytelling and narrative. We seek to problematize the relationship between sound and conflict in theatre practice, to understand the capacity for sound to enhance resilience in participatory musicking[1] and to examine the effects of sound art as a form of socially engaged reconciliatory practice. Looking across different modalities of sound processes and sound production, we compare their significance in multiple fields – drama, informal music making, composition and sonic arts – by assessing the extent to which these are effective media for engaging with processes of resistance, resilience and reconciliation. We then turn to the potential of sound to remediate experience in meaningful ways. We begin, however, by outlining the challenges of analysing sound in research.

Thinking in sound

For many music scholars, the idea of thinking about sound as a distinctive area of inquiry is a fundamental principle. What is the study of music if not, at some level, the study of sounds on the move? The interest in explicating the sounds of music has been spurred in part by the turns beyond the score towards performance and affect evident in the field since the mid-1990s. Likewise, doing anthropology in sound is not new: for example, many scholars point to the work of Steven Feld (1982, 2015) in developing an anthropology of the senses and

---

[1] We follow Small's (1998) definition of musicking, which encompasses any activity relating to the process of music making from listening to performance and the extramusical processes that music making entails.

acoustemology,[2] while in the mid-1980s James Clifford famously asked, 'But what of the ethnographic ear?' in the introduction to his essay collection *Writing Culture* (1986: 12). Yet the broader questions of how to study sound and what sound may convey has taken on a new urgency in a variety of areas since at least the late 1990s. Since then, attention to the role of sound has enlivened discussions of the French countryside (Corbin 1998), pre-historic geography (Mills 2014), jurisprudence (Parker 2015), and commerce (Pinch 2016), to name just a few. Nowhere is this interest more evident than in the rapidly expanding field of sound studies, which challenges both the humanities and the social sciences to attend more carefully to what sounds we hear and how we listen.

Sound studies itself exceeds attempts at definition: within its wide compass fall discussions of acoustics and geography; voice, communication and performance; and media, cultural practices and ontology. Of particular interest is the role of acoustic experience in the formation of the auditory self, described by Steven Connor (1996: 219) as 'a subjectivity organized around the principles of openness, responsiveness and acknowledgement of the world rather than violent alienation from it'. In other words, it is partially through the perception of sound (hearing) that we engage with, or listen to, ourselves and the world. Moreover, within this sound/listening dyad, the link between the *what* and the *how* of hearing is key. As David Novak and Matt Sakakeeny write:

> Sound resides in this feedback loop of materiality and metaphor, infusing words with a diverse spectrum of meanings and interpretations. To engage sound as the interrelation of materiality and metaphor is to show how deeply the apparently separate fields of perception and discourse are entwined in everyday experiences and understandings of sound. (2015: 1)

The dual nature of sound and its consequent impact on scholarly discourse shape the course of research in the sonic arts in multiple ways. Yet for all its theoretical generosity, sound studies has yet to be drawn up fully into the study of conflict transformation. So what does a fuller consideration of sound as both material and metaphor have to offer to the study of conflict and its transformation?

Sound as material and metaphor

From one perspective, thinking about sound as a primary medium is an expansion of analytical potential: no longer constrained by culturally specific

---

[2] Acoustemology refers to modes of sonic knowledge which are particular to place and are acquired by understanding the specifics of a society's auditory culture.

ideas of music, we can explore so-called extramusical or non-musical sounds such as those made by bodies (human or non-human), mechanical actions or electronic manipulation. We might also be more attuned to thinking about acoustic ecologies, or the accumulation of soundscapes. This attunement to the multi-layered sonic character of everyday perceptions and of our discourse opens up new possibilities for cross-genre and cross-discipline interaction. As Canadian composer R. Murray Schafer declaimed in the late 1970s: 'Today all sounds belong to a continuous field of possibilities lying within the comprehensive dominion of music. Behold the new orchestra: the sonic universe!' (1994: 5).

On a more abstract level, conceiving of sound in physical terms as patterns of differential pressure moving through substances insists on a material basis for our analyses. Sound defines, disintegrates and reconfigures the materials through which it passes, and by focusing on sound, we allow for thinking about the sonic properties of spaces, of gestures and of sound's movements over, under, through and around barriers – and conversely, the ways in which sound may be blocked, contained or unheard in certain spaces. Furthermore, sound's materialization of encounter brings attention to the body and might encourage us to follow Roland Barthes ([1972] 1977: 188) in considering 'the body in the voice as it sings', or the voice's 'grain', on equal communicative footing with the semantic content of narrative. In short, sound sutures imagination and communication into the auditory topography of physical bodies and spaces.

Significantly, for the purposes of this study, sound is also fundamentally tied to temporality. Just as it moves through space, sound also moves in and through time and is subject to constant changes and self-transformations as it does so. Since sound is always extending away from its source and is therefore separated from its visual object, we are, in a sense, always hearing a kind of displacement. The transformational potential of time is key to understanding the potential of sonic participatory arts to intervene in formations of resistance, resilience and reconciliation, as well as their remediation in sounding conflict.

Finally, centring sound in our conceptions of participatory art and conflict transformation is an opportunity to pay attention: to attend not only to the ways in which participatory arts engage with and stage the sonic but also to the ways in which they demand listening and evoke a relationship with a listening audience. Building on the distinctions between participatory and presentational styles of music making put forward by Thomas Turino (2008) and theories of relationality by Nicolas Bourriaud (2002) and others, the ways in which sound draws out relationships between resounding objects encourages us to think expansively about the nature of participation. Crucially, this encourages us to reject a hard boundary between producers of sound and hearers of sound, thereby opening up the consideration of different kinds of participatory narrative building in fields as distinct as art music composition, theatre production and storytelling.

Spaces of deafness

In addition to the new experiences it offers, sound contains its own spaces of deafness. For instance, because sound is, as Novak and Sakakeeny (2015: 1) put it, at once perception and discourse, it is tempting to naturalize a single conception of sound as transhistorical and transcultural. This is amply evident in the policing of boundaries between music, sound and their ethically and socially charged counterpart, noise (see Attali 1985). The intrusive but also transformative effects of noise in theatre productions are explored in more detail in Chapter 5. More troubling for the purposes of this volume, it is equally prominent in the still-powerful Romantic idealization of sound as immersive and unifying. When linked to deeply ingrained Western ideals of music (organized sound) as peculiarly powerful for expressing the highest ideals of humanity, the impulse to hype musical sounds as uniquely capable of transforming the divisions of conflict can impede a sober adjudication of their impact. Likewise, Robin James (2019) has recently critiqued a too-easy turn to the sonic as epistemological foundation as merely replacing a contemporary Western visual focus with one that is equally susceptible to the demands of neoliberal capitalism. If by elevating our attention to sound we come to subordinate a qualitative experience to the quantitative demands of market evaluation, any claims to intellectual revolution or reversal of what Jonathan Sterne (2012: 9) calls the 'audiovisual litany' inherited from nineteenth-century conceptions of sound and vision will ring hollow. To guard against such a co-optation, James suggests that scholars must pay attention to the ways in which some aesthetic practices constantly slip out from underneath the demands of capital.

James's description of the ways in which sound may become implicated in the metrics of self-improvement and economic value echo the experiences of many individuals and organizations involved in justifying the role of sound-based creative arts within conflict transformation efforts. We conceptualize sound-based creative arts as arts practices in which both sound and listening are central to the experience, including music and sonic arts, but also theatre, storytelling, speech, poetry and the spoken word. In comparison to the concrete work of establishing legal rights or delivering justice, the capacity of sound-based creative arts to contribute to transforming conflict can sometimes seem as ephemeral as sound itself. Nonetheless, we argue that focusing on the sonic as a significant register has the potential to enhance our understanding of certain kinds of phenomena. The sonic aspects of the arts both convey information and expand the available space for imagination, speculation and transformation. The kinds of knowledge and experience we glean through the cumulative and interactive process of hearing sound, then, are distinct from those we gain through our other senses, and by listening carefully and creatively across genre and discipline we can come to learn about the world and each other in a different way. This is not to imply a simple inversion of the sensorial hierarchy privileging vision laid down in the European Enlightenment but rather to advocate for a theoretical position that attends as closely as possible to the sonic character of our interactions.

Sound and storytelling

Paying attention to sound-in-interaction highlights how storytelling is intertwined with both sound production and with listening. First, storytelling essentially consists of sound making in the form of narratives, and sounds (verbal and non-verbal) create and compose stories. Many conflict transformation projects use storytelling and oral history interviewing as a means not only to give voice to, and thereby approach, experiences of conflict, violence, loss and trauma, but also to address and facilitate processes of resistance, resilience, reconciliation and remediation.

As a process that comprises both listening and speaking, storytelling not only insists on the human right to tell our stories but also emphasizes the ethical duty to hear other people's narratives: to bear witness to their experiences, thoughts, ideas, words and worldviews. As such, it is a process that invites the giving and receiving of empathy with the other's story. Such empathic witnessing can also enable a process of mourning, which creates space for a reconciliation with the loss – and, by extension, also with the other, who is seen as responsible for the loss. In an interview, the French philosopher Paul Ricoeur relates the process of mourning to narration, reconciliation and empathy. Evoking Sigmund Freud's notion of *Trauerarbeit* (or the work of mourning), he explains:

> Th[e] work of mourning is a long and patient travail, which brings under interrogation the ability to narrate it. ... to mourn is to learn to narrate otherwise.

> To narrate otherwise what one had done, what one has suffered, what one has gained and what one has lost. … Freud uses [the] word *Versöhnung*, which has been translated as 'reconciliation.' … *Versöhnung* consists in exchanging roles: each party abandons its claim to be the only one occupying the terrain. Thus, each party must renounce something. (Ricoeur 2005: 23)

The process of mourning opens up narrative possibilities and introduces the principle of plurality. As Ricoeur suggests elsewhere, the ethical importance of storytelling lies in the fact that 'it is always possible to tell in another ways. This exercise of memory is here an exercise in *telling otherwise*' (emphasis in original; 1999: 9). Stories open us up to new perspectives and viewpoints; they introduce new kinds of knowledge and understandings by carrying meaning through emotion. As theatre-maker Anne Bogart (2015) notes, 'Without emotion, you learn nothing.'

Socially engaged arts and the sonic sphere

Many of the practices discussed throughout the book fall under the charged umbrella of socially engaged arts. With its emphasis on process and participation rather than object, the term 'socially engaged' has sparked as many attempts at definition as it has controversy and critique. In one foundational text, Nicolas Bourriaud introduced the idea of 'relational art' in order to package a series of art practices emerging in the 1990s in an aesthetic universe digestible by galleries and museums. From a situation in which community, participatory and local art practices were seen as second-rate and of little consequence to the elite of the art world (ever concerned with object and value) Bourriaud re-framed these social practices in an acceptable theoretical and curatorial discourse. In his 2002 book *Relational Aesthetics*, he writes, 'The possibility of a *relational* art (an art taking as its theoretical horizon the realm of human interactions and its social context, rather than the assertion of an independent and *private* symbolic space), points to a radical upheaval of the aesthetic, cultural and political goals introduced by modern art' (Bourriaud 2002: 14).

At the theoretical core of the relational approach to human interactions and social context is the notion of participation. The participatory ranges from forms of organic collaboration present (if often unnamed) in many artistic practices, to the utopian door to the democratization of the arts and, by consequence, society.

Given the focus on relationship, the dynamics between all agents involved in social practice will determine the mode of operation of a given project. Art institutions, curators, artists, participants and audiences are all together entangled in so-called social engagement, fraught with pitfalls, unpredictability, egos, agenda setting and all that belongs to the human condition. As such, it is unrealistic if not delusional to think of participation as a liberating and democratizing strategy in and of itself.

This scepticism towards framing participation as inherently democratic extends to other relational spaces, including the tension between the often-inflated rhetoric used in project communication and curatorial discourse and the realities of participation and the possibilities of evaluation or reflection. The intentions of artists involved in socially engaged work are often tied into ostensively positive, transformational and liberating impacts on the participants themselves rather than on 'external' viewers or listeners. Much of the artistic value, if one wants to put it that way, resides within the projects themselves. It is the inner logic and relational dynamics of each project and its participants that is the art. As a result, external presentation of the process is often fraught with difficulties as is a project's evaluation and critique. The arguably direct mode of access to artistic practice through the viewing of an object or the act of listening to a piece of music is here replaced by myriad threads of activity, diffused authorship and fragmentary media which are often only brought together through a post-participation constructed narrative. The issues around the construction of this narrative (who constructs it and who is it for) is at the centre of Claire Bishop's critique of socially engaged art and participation in *Artificial Hells*:

> Today's participatory art is often at pains to emphasise process over a definitive image, concept or object. It tends to value what is invisible: a group dynamic, a social situation, a change of energy, a raised consciousness. As a result, it is an art dependent on first-hand experience, and preferably over a long duration (days, months, or even years). Very few observers are in a position to take such an overview of long-term participatory projects: students and researchers are usually reliant on accounts provided by the artists, the curator, a handful of assistants, and if they are lucky, maybe some of the participants. (2012: 6)

Bishop's critique also points to the importance of community in discussions of participatory art. Whether pre-formed or project specific, organic or artificial, the sense of community in socially engaged practices is often valued and nurtured. It provides focus and a sense of purpose, and it is a vehicle for developing perhaps the most important characteristic of socially engaged practice: trust. Yet despite its importance, it is an ephemeral – and possibly chimeric – characteristic.

As the recourse to Bourriaud (2002) and Bishop (2012) suggests, much of the criticism surrounding socially engaged art practices derives from visual art aesthetics and discourse. In particular, the influence of Bourriaud's relational aesthetics in curatorial practices or Bishop's concern with the spectacle and spectatorship (or lack thereof) position socially engaged work in the lineage of contemporary museum and gallery practices. Nonetheless, the substance of these critiques extends beyond the walls of the museum or gallery. What is contested is the object-based affordance for critique by the art world (in whatever media it might manifest) to an art practice that aims to 'do', to change and to transform social conditions. In one of the first articulations of this debate Suzanne Lacy writes, 'The underlying aversion to art that claims to "do" something, that does not subordinate function to craft, presents a resonant dilemma for new genre public artists' (1995: 20). Twenty-five years on from Lacy, this 'resonant dilemma' continues to shape the socially engaged sonic arts. Throughout the book, the term 'socially engaged arts' captures a wide variety of practices, all involving participation, and, to different extents, conflict transformation. Each practice, institution and project deals with the characteristics discussed above in their own unique ways, often in the context of the very specific conditions under which they operate.

Sound, word and interpretation

As we have outlined above, the focus of this book is not so much on what sounds – whether verbal or non-verbal – do but on *how* they operate and are employed to strategic effect in protracted conflict settings or in regions recovering from conflict. We are especially concerned with the potentialities of sound to contribute to the work of peacebuilding. Our emphasis on sound in this analysis conjoins what are often considered in the West to be separate aesthetic domains of the audio, linguistic and visual. Instead, as we shall see in these creative performance domains, sound, word and image coalesce in a flow that variously comprises discourse, practice and embodiment, a confluence that is fundamental to the work of peacebuilding through the arts. Thus, our analysis of sound's relationship to meaning making is not constrained to a static

Western abstraction of text or discourse alone, but rather the text, the sound and the speech event are understood as 'interlac[ing] with the moment, actively influencing it' (Kersenboom 1995: 16). The same is true in contexts of musicking whereby musicians affect their surroundings in the process of playing, that is, through 'sounding out interpretation[s]'.

One term that has frequently been applied to this sense of multimodal sonic flow is that of 'soundscape', which was popularized by Schafer ([1977] 1994) and has since become a widely recognized (though frequently controversial) concept with significant prominence for policymakers. As such, it is a useful heuristic for communicating with non-academic audiences and for activating policy change. Nonetheless, in what follows, we avoid using the term which has at times been criticized for being too general to do the work required of it for the particularities of musical and performative interaction. By situating sound at the centre of this work without glossing everything as 'soundscape', we give prominence to an otherwise marginalized perspective that is frequently conflated with the arts, and one which in turn is often viewed as a 'soft' dimension to peacebuilding in contrast to the 'hard' questions of conflict that are more often dealt with by politics and the social sciences (Shank and Schirch 2008: 217).

We also recognize the entangled, ambiguous and, at times, conflictual connotations that can arise from concepts of sound, sounding and the sonic. Nevertheless, we adopt these concepts to highlight, elaborate, reveal and deconstruct an understanding of 'deep and thick listening experiences' (Schulze 2021: 11) and how they are variously influenced by the conflict contexts that shape them. This approach draws together work in the anthropology of sound (Feld 2015) and the senses (Howes 2005; Howes and Classen 2013) as it addresses questions of how modes of resistance, resilience, reconciliation and remediation can be experienced through 'sounding as' and 'sounding through knowing' which in turn shape modes of listening and feeling (Feld and Brenneis 2004). The listening we address is a form of attunement not just to words and gestures but to sensing modes of communication and reception through performative collaboration. Listening mediates internal responses to others and thus shapes the potential for action. As Salomé Voegelin (2011: 108) notes, 'Sound binds us to the world as into an indivisible cosmos – an invisible and indissoluble volume of interbeing – where the meeting of our sounds performs a cultural topography of coexistences, a voluminous map of our asymmetrical being together.' In our analyses, we seek to understand the complex affectivities that arise from these processes of sounding and listening in 'interbeing' and how expressions of resistance, resilience and reconciliation are (re)mediated by engaging with sound and listener responses to staged and participatory performances.

Exploring resistance, resilience, reconciliation and remediation

The interplay between our core concepts of resistance, resilience, reconciliation and remediation enables us to engage firstly in a series of critical debates about how individuals and groups can 'mobilise and strategise their effects in the sounds and performances of everyday life' (Magowan and Donnan 2019: 1). Secondly, we examine theories of sound through socially engaged arts' practices to illustrate how creative resonances variously pervade sociopolitical imaginaries of conflict (see Daughtry 2015) and can produce beneficial somatic effects in creative practice (see Abrahams and Lily 2015; Levine 2010). This introductory chapter takes each of our theoretical frames separately, working through critical approaches to the concepts, analysing their applications within sound and music and foregrounding potential impacts which will be further explored in the case studies in the following chapters.

Issues of sound and its expression in drama, music making and storytelling frame the discussion of the orienting lenses that define each chapter of the volume. In each case, this book examines how individuals and communities respond to violence and forge paths towards peacebuilding through the creative potentialities of listening to sound within and across socially engaged participatory arenas. Each of the later chapters in the volume takes as its focus one or more fieldwork areas, the Middle East (Lebanon and Palestine), Northern Ireland and Brazil, alongside one chapter focusing on how the displacement of people by conflict has created new sonic environments for refugees. This reveals the multitude of ways by which social relations are ecologically and performatively interwoven in an acoustemology of distinction and interconnectedness. Through case studies undertaken collaboratively with partners, we explore how an intertwined framework of sound and participatory practice generates innovations in listening, that is, alternative responses to the effects of conflict and embodied processes of personal, social and, in some cases, political transformation.

Each of these case studies highlights how participatory arts are employed by artists, musicians, facilitators, theatre practitioners, community activists and other stakeholders as a means of 'strategic creativity', with a range of aims, including transforming trauma, promoting healing and facilitating empowerment. This, then, helps us reflect on the multifaceted nature of resistance, resilience, reconciliation and remediation in sound and conflict. In the sections that follow, we contextualize each of these key terms within academic literature and creative practice.

Resistance and creative practice

Resistance is a persistent theme during conflict, and its meaning varies widely in theory and practice. It can originate at the individual, community or institutional levels, and likewise, it can target individuals, institutions, governments or social structures. Resistance can be spontaneous, tightly coordinated or anything in between, and it can be covert, overt or a mix of the two. It can take the form of violent actions, protests, demonstrations, marches, vigils, strikes, boycotts, civil disobedience, sabotage, graffiti, music, writing, art and a myriad of other tactics. Moreover, resistance can be an integral part of the conflict, a response to the conflict or a personal or communal way of transforming or transcending the conflict. As Hollander and Einwohner (2004: 538) note, despite its broad scope, resistance typically includes a sense of *action* and a sense of *opposition*. It is active in that it is more than a state of being; actors and activists who resist engage in some sort of behaviour, however subtle. Similarly, that action 'occurs in opposition to someone or something else' (539). A relatively comprehensive definition of resistance integrates both of these attributes: resistance is an action that challenges someone or something else. We recognize that other definitions may be more apt in other contexts, but we find this definition most applicable to the themes and ideas we are covering in this book.

Resistance movements often overlay with creative practices. As Emery (1993: 347) writes, the 'artistic activity for inner resistance to group regression and tyranny' is evident throughout history and philosophy; Rousseau regularly 'contrast[ed] his own creative self with the corruption of the surrounding social order', and John Stuart Mill emphasized 'artistic individuation as a key check on the tendency to conformity of modern mass society' (347). Since the late twentieth century, creative practice in resistance movements have included all mediums, ranging from the writings and essays of Vaclav Havel in Czechoslovakia during the Velvet Revolution; to the protest music of the 1960s (Heilbronner 2016); to the *arpilleras*, or narrative quilt squares sewn by Chilean women to protest injustices under Pinochet's regime in the 1970s (Onion 2014); to the graffiti and street art in Palestine during the First Intifada (Peteet 1996); and, in recent years, to the digital photography and videos that have documented protests around the world in contexts as varied as Hong Kong, South Sudan, Iran and the United States.

Sonic practices, including music, storytelling and theatre, have long resonated in resistance practices. In conflict contexts in particular, sound-based approaches can be amplified or minimized as necessary. On the one hand, sonic practices may allow for performances to be recorded, enabling replication and ease of distribution, as with protest songs, hip-hop tracks (Martinez 1997) or carioca funk (Palombini 2013). In other contexts, a sound-based performance (without recording) can be strategic in terms of leaving no tangible evidence that might be used against activists by the state or regime. Sonic practices of resistance can be individual or collective and can be organized or spontaneous. They can include voice, instruments, impromptu instruments (e.g. in the Latin American tradition of *cacerolazo*, or banging pots and pans) or even silence (e.g. Turkey's 'standing man' protest in 2013). Likewise, they can create artistic practices that become appropriated (often by the very power that is being resisted).

In this project, we focus on socially engaged sonic practices, which 'have a primary interest in participation, affecting social dynamics, dialogue, and at times, political activism' (Rebelo and Cicchelli Velloso 2018: 137). Given that these highlight the participation of community members, the focus is not solely on the artwork, but on the production process itself. As our work demonstrates, this can be valuable for galvanizing resistance in terms of facilitating personal expression, interpersonal exchange and community organizing and cohesion.

Resilience through musicking

Although the concept of 'resilience' has proliferated in recent years, critical analysis of the term highlights a certain misrecognition of its significance. This misrecognition can be characterized by an emphasis on individual responsibility (echoing some of the tenets of neoliberal ideology), as distinct from the emergence of resilience in ecological studies and systems theory. These out-workings of neoliberal and ecological conceptions of resilience are, in many cases, diametrically opposed. Such tension between individualized and ecological or collective/societal understandings of resilience is evident in our case studies.

Resilience as a systems theory concept is widely used to refer to the capacity of systems to withstand pressure and avoid complete breakdown, while also reflecting upon the impact of stressors in order to ameliorate their effects in the future. Resilience underpins all aspects of society but has become more prevalent in ecological discourse as climate emergencies, in-country catastrophes and conflicts lead increasingly to instability in social systems. At the core of resilience is a complex paradox between stability and disturbance that requires a move away from dominant systems' attitudes to embrace resilience as 'a disruptive concept: one that calls for radical transformations' (Manzini and Till 2015: 10). Nowhere are such radical transformations more prominent than in the arts. However, whenever we speak of resilience in the arts, a different set of criteria applies to that which might be considered necessary to rebuild resilience in a city's infrastructure as architectural product. Instead, resilience as a process engages qualities of cooperation, self-care and care for others, generosity, kindness, trust, empathy, and reciprocity, among other elements.

While there is now a growing field of the arts and conflict transformation and music in peacebuilding, much of the literature relating to these processes of resilience examines how music supports the recovery and rehabilitation of those who have suffered from the effects of war and violence. Analyses of cases of recovery tend to stress the significance of individual healing (Austin 2002), even though music is also used in the healing of societies. Those who adopt this perspective stress the effectiveness of listening and musicking in providing solace, empathy and spirituality, whereby music is able to guide empathetic relationality and enhance capacity not only for resilience but also for peacebuilding (Laurence 2008; Urbain 2008a). Music can be a versatile tool in its capacity to build bridges between those willing to view the past and their distinct cultural traditions in a shared light, and thus, it offers hope for the potential for reconciliation. It can also facilitate the dissolving of anxieties and tensions and open the capacity for non-verbal dialogue that involves both sound making and listening. For example, a Norwegian school music education programme brought together pupils from immigrant groups to play together using their own musical traditions, helping 'to reduce ethnic conflicts and interracial tension among pupils, fostering empathy, improving social relationships and strengthening the self-image of the immigrant pupils' (Skyllstad 1997, 2008; cited in Grant et al. 2010: 191).

In this book, we employ the term 'resilience' to examine how sound, music and movement can be used to address the effects of violence and enhance a participant's capacity to manage disturbances, increase self-awareness and self-management, and become flexible in their approach. Dillon (2011: 240) argues that immersion in music making is akin to the concept of resilience, as it needs to be 'personal, social and cultural' and each 'build[s] competencies that act as protective factors with positive implications in other aspects of life'. The questions that we are dealing with in this volume follow on from Dillon's observation to ask how do sound, music and movement enhance interpersonal competencies; in what ways are performances meaningful encounters with others; how can these interactions be measured and assessed and to what extent do they lead to changes in participants' social and cognitive engagements?

The performative dimensions of reconciliation

At least since South Africa's Truth and Reconciliation Commission, 'reconciliation' has emerged as a hegemonic concept governing political transition after violent conflict. In his IDEA Handbook on conflict resolution, David Bloomfield and colleagues argue that it needs to be understood as both a goal and 'an overarching process which includes the search for truth, justice, forgiveness, healing and so on' (Bloomfield et al. 2003: 12). Yet, despite its popularity and generally acknowledged significance, 'reconciliation' remains a contested and diffuse term that generates multiple meanings in different contexts. The scope of meanings is evident in the images that are associated with the term, which range from gestures, such as platonic embraces and handshakes, to the symbolism of bridging a divide, invoking an attempted coming together. In other words, it always concerns a *relationship* between two or more entities (see also Bloomfield 2006: 8; Lederach 2001: 842). As Cohen (2006: 74) suggests, processes of reconciliation involve 'former enemies acknowledging each other's humanity, empathizing with each other's suffering, addressing and redressing past injustices, and sometimes expressing remorse, granting forgiveness, and offering reparations'. Interestingly, this formulations thereby also suggests an aural spectrum from silent emphatic acknowledgement to audible verbalization of responsibility.

Reconciliation is suspended between the temporal coordinates of the past and the future, as reflected in the following two definitions from the *Oxford English Dictionary* (OED):

1. b. The action of restoring estranged people or parties to friendship, the result of this and the fact of being reconciled.
4. a. The action or an act of bringing a thing or things to agreement, concord or harmony – the fact of being made consistent or compatible.

Figure 0.1 left, Josefina de Vasconcellos's 'Reconciliation' (Stormont Castle, Belfast); right, Maurice Harron's 'Hands across the Divide' (Craigavon Bridge, Derry).
Source: Photos by Matthew Logue.

The first definition is directed to the past, evoking a return to a prior state of friendly relations – in other words, a reunion. This conception is visualized by Josefina de Vasconcellos's 1977 sculpture – tellingly entitled 'Reunion'; it depicts a man and woman embracing each other. In 1995, a bronze cast was made to mark the fiftieth anniversary of VJ Day. Renamed 'Reconciliation', it was erected in the grounds of Stormont Castle, Belfast (see Figure 0.1). The way in which 'Reunion' became 'Reconciliation' tells us something about a specific understanding of reconciliation that is based on the notion of a pre-existing state of unity and harmony. But when we look at the context of many post-conflict societies, this is problematic. As Antjie Krog notes in relation to South Africa, there is 'nothing to go back to, no previous state or relationship one would wish to restore' (1999: 108). Nonetheless, this implied focus on the past emphasizes the legacy issues that continue to haunt reconciliation processes.

In turn, the second definition of reconciliation is directed towards the future, focusing on an anticipated goal through a process that starts in the present and is, literally, in the 'making'. An apt visual analogue is the famous statue 'Hands across the Divide' situated at the west end of Craigavon Bridge in Derry/Londonderry: each figure reaches out a hand toward the other, but they do not yet touch. The sculpture was produced by Maurice Harron in 1992, on the twentieth anniversary of Bloody Sunday. In a speech in 1995, Bill Clinton captured the sculpture's anticipatory dimensions: 'It is a beautiful and powerful symbol of where many people stand today in this great land. Let it now point people to the handshake of reconciliation. Life cannot be lived with the stillness of statues. Life must go on. The hands must come closer together or drift further apart' (Clinton 1995). With hindsight, Clinton's statement seems prophetic about the continuous and precarious unfolding of the peace process, but it also highlights the doubleness within the concept of reconciliation, which both looks to the past to redeem the present and future and anticipates a better future.

Remediation, the past and the future

Our final key term, 'remediation', is also intimately connected to the unfolding of time. It is also a term with a dual life in media studies and in cultural studies. In the former, remediation is understood as the dynamic process by which one medium engages with or appropriates another, as when a film is based on a play, when a hand-drawn illustration is replaced by a digital one or even when a museum represents its collection of physical objects through a curated series of photographs on their website. This transformation of forms is a long-standing feature of media culture, but the turn to digital formats within the so-called new media has facilitated these processes of borrowing and reworking (see Bolter and Grusin 1999). In contrast, from the perspective of cultural studies, remediation is thought of as an intermedial reiteration that lays the foundations for collective understandings of the past and present (see Erll and Rigney 2009). This foregrounds the ways in which different media forms circulate and – through that circulation – reform our consensual understanding of reality.

At the intersection of these academic applications, remediation emerges as the creative reiteration and recombination of media forms as a means of resituating the past. Yet as Peter McMurray (2021) points out, remediation also contains within it the much older etymological resonance of healing, or of 'remedy'. To remediate is to reverse damage or to correct something that is deficient through a restorative process. Like reconciliation, then, it looks both backwards and forwards in time in pursuit of transformation.

In this volume, we synthesize all of these uses of remediation in order to examine how peace workers, performers, composers, theatre practitioners and others transform sounds and stories across different media contexts, how they carry and disseminate these stories across time and space and how such sonic remediations of experience contribute to resistance, resilience and reconciliation. In short, we suggest that sonic remediation is not simply a question of tracing the intermedial transformations of sound as abstract phenomena but rather of paying attention to the ways in which participants in these musical encounters use sound to structure sociality as a potential step towards remedying damaged and broken relationships. Crucially, we explore these dynamics through the lived experiences of individuals and groups who use sound-based practices as interventions to build peace and push for social change in their communities.

Case studies

In the chapters that follow, we explore our collective methodologies (Chapter 1) before exploring each of our case studies in detail. Chapters 2 through to 5 each focus on one key theme (resistance, resilience, remediation and reconciliation), while Chapter 6 details the creation and performance of a sound art installation that draws on the body of research that animates the *Sounding Conflict* project as a whole. This is followed by a short Conclusion, in which we draw together the key threads of the case studies and present an overview of the work of sound, music and storytelling as important tools of resistance, resilience, reconciliation and remediation in creative practice.

Resistance: Performing the frontline

Chapter 2 focuses on case studies of two community partners in Lebanon. We chose Lebanon because it has experienced various cycles of conflict, having rebuilt gradually from a civil war that lasted from 1975 until 1990 while still harbouring many sectarian divisions. Lebanon also experienced a social movement or 'revolution' in 2019–20. Protests against a government seen as being dysfunctional and corrupt have continued to the time of this writing. Lebanon also complements other key case studies in the book, specifically Northern Ireland; both are 'post-conflict' societies still largely defined by secular divides and tensions. Also, like Northern Ireland, there have been no formal mechanisms for dealing with the past in Lebanon, with most reconciliation and accountability initiatives coordinated by civil society rather than the state. At the same time, these grassroots efforts have yielded creative responses to confronting both the earlier conflict and present divisions, warranting closer research of arts- and sound-based practices.

In Lebanon, we worked with Fighters for Peace (FFP), an NGO comprised of ex-combatants from Lebanon's civil war that formed in 2013 to try to prevent present-day violence along sectarian lines by sharing their stories and experiences of war through dialogue sessions, documentary films and video testimonies.[3] We also worked with Laban, a performance art theatre in Beirut. Laban focuses on using improvisational theatre to inform civil society activism through interactive performances such as Forum Theatre, in which a play is used to open audience discussion on difficult issues, and Playback Theatre, in which actors creatively interpret an audience member's story or experience to open dialogue on challenging topics. Laban has also worked closely with ex-combatants from FFP to use theatre as a creative means for starting difficult conversations about accountability and reconciliation.

In post–civil war Lebanon, 'memories of the war were effectively silenced, giving rise to an overdetermined collective amnesia informed by political and commercial expediency' (Launchbury et al. 2014: 457). Through storytelling and theatre, respectively, FFP and Laban resisted this collective post-conflict silence, not to reignite the tensions of the civil war but to deal with the individual and collective impacts of the past. During the 2019–20 revolution, members of both FFP and Laban extended this resistance by challenging the state in the public space of protest. The participatory nature of their activities made them accessible to the broader public, facilitating an open space for critical dialogue and discussion among other activists and community members. These forums provided opportunities for cultivating individual and communal resilience to state crackdowns of protesters on the one hand and pressures from violent agitators on the other (Collard 2019). FFP's dialogue sessions and Laban's post-performance discussions also offered spaces for reconciliation, or at least enhanced understanding by including participants from different backgrounds often characterized by divisions and suspicion. These divides are most notable between sects, but gaps also persist between religious and secular, urban and rural, and rich and poor. FFP was also instrumental in bridging intergenerational divides during the revolution, with some youth initially distrustful of the older generation's willingness to push for change and accountability (interview with author). FFP's participatory approach and use of storytelling and dialogue enabled them to counter that assumption and deepen solidarity networks between older and younger activists during the uprising.

For FFP, individual ex-combatants went through a long self-reflection process, confronting their own role in the conflict and speaking about it openly even when pressured to stay silent. Likewise, Laban provided another venue for the

---

[3] Julie Norman is an academic advisor and former board member for FFP.

ex-fighters to reckon with their pasts in a non-confrontational way, while also allowing the younger actors to better understand and express the intergenerational trauma (Shaar 2013) stemming from the conflict. Both also engaged others in these processes of reflection: FFP through their dialogue sessions with youth and community members and Laban through their interactive theatre techniques and post-performance discussions. These socially engaged creative practices enabled both FFP and Laban to explore often-taboo topics and draw others into conversation, building resilience and, at times, reconciliation.

Resilience in creative practice: Musicians Without Borders

From the theme of resistance, in Chapter 2 we turn to examining creativity through an approach to nonviolence in two programmes run by Musicians Without Borders (MWB), an international charity organization that uses music for peacebuilding and social change. In collaboration with local musicians and organizations, they 'bring music to people and places affected by war, armed conflict, and displacement'.[4] We focus on just two countries in which they operate: Northern Ireland and Palestine. In Palestine MWB is located near Bethlehem in the West Bank of the Palestinian territories and works primarily with Palestinian refugee youth.

In a different context, in Northern Ireland, MWB ran a three-year programme from 2014 to 2017 in collaboration with the Derry/Londonderry cultural organization Cultúrlann Uí Chanáin for participants who were training to be accredited facilitators with the organization. Participants came from a variety of backgrounds and regions and further aimed to use the skills they had learned in their own communities to effect change with the music and drama groups they were leading. This second case study was located in Derry/Londonderry, where a history of conflict has impacted the city and its surrounds. However, the musical ethos of MWB is not to address violence per se in these workshops but to start from a principle of nonviolence by emphasizing the key criteria of 'safety, inclusion, equality, creativity and quality'.[5] For example, the use of instrumentation that does not require specialist training creates a field of practice in which musicians' skills

---

[4] Available online: https://www.musicianswithoutborders.org/?gclid=CjwKCAjw-8qVBhANEiwAfj XLriz1ZKgHJuOu2WrGFeH4hNLg0V9Stk-CQBW4AuL-OHZInn8wUTU1IhoCbZwQAvD_BwE (accessed 8 September 2022).

[5] Available online: https://www.musicianswithoutborders.org/how-we-work/music-leadership/ (accessed 8 September 2022).

are perceived to be equally expressive in MWB workshops. This democratizing effect can further be achieved through participatory song writing where all participants in the group are invited to have equal input and to generate shared creativity. The effect of this process is to consolidate 'common ground' (Dillon 2007) through 'delineated meaning' (Green 2008) as participants strengthen their bonds of connectedness and persist in their efforts to achieve a positive outcome. Further, MWB's approach resonates with that of music therapists who have found that establishing a safe space can alleviate traumatic experiences brought out by music and increase tolerance towards these experiences (Bensimon et al. 2008). The effects of participatory music making can generate shared identities, strong interconnections between participants and a sense of achievement and joy that increases a hopeful resilience.

Each of these contexts illustrates very different approaches to the use of music. We consider how music facilitators employ different kinds of sound objects and vocalized rhythmic activities to bring together three specific aspects – 'acoustical sources, inhabited space and the linked pair of sound perception and sound action' (Augoyard and Torgue 2005: 7) – through immersion in their respective musical practices. We further explore how these sonic effects are core to perceiving and interpreting the creative environment as well as informing participants' musical histories. Through their respective forms of musicking in Palestine and Northern Ireland, we show how different psychosociological dynamics become intertwined with the 'sound marking of inhabited or frequented space; sound encoding of interpersonal relations; symbolic meaning and value linked to everyday sound perceptions and actions; and interaction between heard sounds and produced sounds' (8). It should be noted that sonic effects allude to a much wider elaboration of the sociocultural milieu in which they exist, as they shape the internal responses of individuals as well as wider social relations. In this sonic nexus, we explore how participatory musicking generates relations of trust that are foundational to resilient interactions.

Remediating refugee stories

This is followed in Chapter 4 by a case study of a very different sonic (and geographical) context that extends our exploration into new kinds of formalized sonic practices. Although the capacity for creative arts to engage in narratives of resistance, resilience and reconciliation is often studied *in situ* in 'post-conflict' societies such as Northern Ireland or Lebanon, the disruptive nature of conflict means that those affected by it are often displaced into new societies as refugees or forced migrants. This creates new challenges for listening to, imagining and narrating relationships, and this case study examines the remediation of refugee stories through a close examination of individual stories told through contemporary avant-garde musical compositions.

The focus of Chapter 4 is on two compositions by the composer and documentarian Mary Kouyoumdjian (b. 1983): *Paper Pianos: 'You Are not a Kid'* (2016), for small instrumental ensemble, voice and electronic playback, and *They Would Only Walk* (2020), for amplified string quintet, string orchestra and audio playback. The title of *Paper Pianos: 'You Are not a Kid'* focuses attention on the relationships between sound, music and media, which are drawn out in the story it tells of Milad Yousufi, a young man born in Afghanistan in 1995. As a child, Milad was fascinated by music, but living under Taliban rule meant that acquiring an instrument proved to be impossible. In its place, Milad resorted to drawing out the pattern of a piano keyboard on pieces of paper. He taught himself to play soundlessly on this paper piano and after the Taliban's fall from power took up formal piano training in Kabul. Only a few years later, as the conflict in Afghanistan increased once more, Milad fled his home, ending up as a refugee in New York City in 2015, where – on the hunt for an actual piano on which to practice – he was introduced to Kouyoumdjian. Milad's story of his childhood in Afghanistan forms the backbone of *Paper Pianos: 'You Are not a Kid'* via the inclusion of recordings taken from Kouyoumdjian's interviews.

In contrast to the intermingled sounds (and silences) of war and music in *Paper Pianos: 'You Are not a Kid'*, *They Would Only Walk* focuses on flight and refuge. Like the earlier piece, it gains its title from its constituent text, this time a story told by Yamilette García about how her father crossed the border from Mexico to the United States on foot. In this piece, Yamilette's words are interwoven with other stories told by refugees and forced migrants who have arrived in the United States from around the world. The commission originated from an invitation for Kouyoumdjian to be the composer-in-residence at the Buffalo String Works (BSW) in Buffalo, New York, in 2019. BSW is an organization dedicated to providing high-quality musical education to refugees, immigrants and others as a means of 'cultivat[ing] youth to be agents of change' in the city and beyond.[6] The piece was premiered by BSW's own students together with members of the Buffalo Chamber Players in a live-streamed

---

[6] Available online: https://www.buffalostringworks.org (accessed 12 September 2022).

benefit concert in March 2021. Taken together, these two compositions and the interviews conducted with their participants provide excellent opportunities for investigating how individuals and communities negotiate the possibilities of remediating relationships between refugees, their families and the wider society.

From noises of conflict to dissonant sounds of reconciliation

In Chapter 5 we return to Northern Ireland, where, in face of the absence, or the contentiousness, of an overarching institutional framework for dealing with the legacy of the past, reconciliation endeavours have been largely devolved to independent initiatives, community projects and especially the arts. This has resulted in widespread funding for arts and especially theatre projects that promote reconciliation and healing, such as through the PEACE IV Programme. The focus of Chapter 5 is on the sounds used in four recent plays by three theatre companies based in Belfast, Northern Ireland: Kabosh, Tinderbox and TheatreofplucK. The chapter proposes that noises in these plays are associated with conflict, which can attain a transformative power that enables new perspectives and meaning making to emerge, which, in turn, can open new pathways to approach reconciliation.

Founded in 1994, Kabosh is dedicated to producing 'provocative theatre that transforms our understanding of who and where we are, through giving voice to site, space and people'.[7] It has been under the artistic directorship of Paula McFetridge since 2006. The two productions that are examined here specifically deal with the ongoing legacy of the Northern Irish conflict. Laurence McKeown's *Those You Pass on the Street* (2014) was commissioned by Healing Through Remembering, an independent organization focused on finding ways to deal with Northern Ireland's past, 'to assist different communities examine the difficult subject of reconciliation' by exploring the entanglement of family ties in issues of forgiveness and forgetting.[8] McKeown's *Green and Blue* (2016) is based on testimonies from the Royal Ulster Constabulary and Garda Síochána officers who served on the Irish border during the three-and-half decades of violent conflict, euphemistically known as 'The Troubles'. The play explores the theme of friendship across the divide. Both productions use intrusive noises to alert us to the potentiality of conflict that can be transformed by attentive listening and meaningful engagement with 'the other' as putative opponent.

---

[7] Available online: https://kabosh.net (accessed 8 September 2022).
[8] Available online: https://kabosh.net/what-we-do/legacy-of-conflict/ (accessed 8 September 2022).

Tinderbox Theatre Company was founded in 1988 to create 'challenging theatre not ordinarily seen in Belfast'.[9] The company has been dedicated to new writing by upcoming playwrights, and under the artistic directorship of Patrick J O'Reilly since 2016, it has been 'inspired by the ever-changing creative landscape of new contemporary arts practice, new and ambitious theatrical forms and the rich diversity of artists and participants'.[10] This case study focuses on Tinderbox's 2019 adaption of Alfred Jarry's *Ubu Roi* as *Ubu the King*. Originally founded by Dr Niall Rae and colleagues in 1998 in Philadelphia, TheatreofplucK became Northern Ireland's first publicly funded queer theatre company, exploring specifically issues of lesbian, gay, bisexual and transgender identities in Ireland. Here, we examine TheatreofplucK's 2018 performative verbatim audio walk, *So I Can Breathe This Air*, written by Shannon Yee and based on interviews with the Rainbow Project's Gay Ethnic Group (GEG), which was originally entitled *Multiple Journeys (of Belonging)* (2017). In very different yet comparable ways, sound in both the Tinderbox and the TheatreofplucK production has a powerful transformative effect which is experienced as distressing and disturbing, as well as empowering and liberating. If *Ubu the King* immerses us into the soundings of war and conflict, *So I Can Breathe This Air* creates a space for performing reconciliation. All four productions offer indicative case studies for exploring the ways in which theatre performances can, through the use of sound effects, offer new approaches and understandings to reconciliation.

Working through creative practice

In Chapter 6, we present 'Sounding Conflict: A Performance in Five Acts'. The artwork was conceived in dialogue with the case studies outlined above and expanded on through the various chapters in the book. The work resulted from a collaboration between Pedro Rebelo, Matilde Meireles and Tinderbox director Patrick O'Reilly. Within a framework of socially engaged arts practice (Lacy 1995; Rebelo 2008; Helguera 2011), the work takes a primarily non-verbal approach to exploring how resistance, resilience, reconciliation and remediation manifest themselves at various stages of conflict and across different sociocultural contexts. Taking the building, destruction and rebuilding of walls – manifested

---

[9] Available online: https://www.culturenorthernireland.org/article/900/tinderbox- (accessed 10 November 2021).
[10] Available online: https://www.tinderbox.org.uk/about-us/ (accessed 1 September 2021).

here both representationally and sculpturally with wooden bricks and sand – the work is a cyclical performance piece which responds to both broad and specific insights from the fieldwork.

Sonically, the piece is composed of fragments drawn from field recordings, urban and rural environments and specific 'earmarks', together with sound design and a hip hop montage. Throughout the case studies that make up the Sounding Conflict project, hip hop emerged as a common musical practice engaging with conflict and community. In the artwork, Jordanian hip hop and rap (Milton-Edwards 2021), Irish-language rap (Ó hÍr 2021) and the rich tradition of carioca funk in Brazil (Palombini 2013) are blended and juxtaposed in performative constructions and deconstructions. The work is presented as a film that depicts two characters building and rebuilding walls in five visually similar, but sonically and emotionally distinct, scenes or acts. Each act is framed by a quotation taken from field research and sets the tone for an open interpretation of the actions on screen, at times direct but often ambiguous. The wall as metaphor is explored to shift its meaning from barrier to protection. Questions around who builds walls and borders and for whom are raised by the very performance actions we see on screen. Moreover, the film is intended to be experienced in an immersive context through projection mapping onto sculptural elements in a gallery context and spatial audio, inviting the audience to momentarily inhabit the world of the performance.

Chapter 6 outlines the creative strategies behind the artwork while investigating a sonic understanding of resistance, resilience and reconciliation based on fieldwork conducted in the Som da Maré project in Rio de Janeiro, Brazil, in 2014. The creation and production of 'Sounding Conflict: A Performance in Five Acts' represents a type of creative practice informed by other research methodologies but ultimately aiming to articulate realities and conditions often difficult to access. By leaving the interpretative act to the audience, we invite active listening and embodied engagement. In an aesthetically framed context, the work attempts to expose complexities, contradictions and ambiguities of the various conflict conditions addressed throughout this book.

Conclusion: Sounding out conflict in the sonic arts

Arts-based creative approaches such as these we study here have become increasingly popular. Moreover, they have proven to be an effective means to initiate, support and maintain lasting conflict transformation processes in tandem with sociopolitical and economic structural changes. Yet this does not mean that they present a single or unified field of study. In this, they reflect some of the difficulties attendant on the widespread use of the metaphor of harmony as the end goal for social and cultural transformation (see Korum 2020). In particular, the simple equation of sociocultural harmony with consonance imposes a Western-centric tonal understanding of musical sound onto social organization, while, as Geoffrey Baker (2014: 208–9) points out, an emphasis on unity devalues the creative potential of dissensus. In response to this problematic, Solveig Korum (2020: 55) suggests the notion of 'dissonant harmony' as a more productive metaphor, and Gillian Howell (2018) has suggested a framework that distinguishes five categories of (dis)harmony. While we do not employ Howell's framework, her discussion of multiple interlocking harmonies offers a useful analogue to our interweaving of resistance, resilience, reconciliation and remediation. In each of our case studies, these processes underscore the multifaceted and interrelated nature of those concepts at work in conflict and post-conflict contexts. This was evident in the individual experiences of the activists, artists and performers producing the sonic practices, as well as the collective experiences of the audiences, participants and communities listening and responding to them. By taking this more conflicted and conflictual approach, we resist the romanticization of harmonies both methodological and sociocultural in favour of a nuanced exploration of sound and conflict.

The complex interweaving of experience is likewise evident within the fabric of this volume. In writing this book, we have brought together several different approaches to the sonic arts, in that we all listen out from our respective disciplines. Several of the contributors (Magowan, Rebelo and Phillips-Hutton) listen out from music studies, though there are distinctive inflections of anthropology/ethnomusicology (Magowan), composition/performance (Rebelo) and musicology/philosophy (Phillips-Hutton). One contributor (Norman) listens from the discipline of political science, while still another (Lehner) listens from those of drama and English. Each of these listenings refracts the sonic through distinct lenses, resulting in what Michael Bull (2021: 18) has termed a 'prismatic' approach to genres from music, to politics to theatre. Yet tracing the role of sound through this kind of juxtapositional thinking allows us to map the sometimes surprising character of sounds and the relations they instantiate across different artistic forms.

This interdisciplinarity, in turn, helps us bridge a methodological gap that exists between what Gallagher and Prior (2014: 272) call 'sonic ethnographies' and what they call 'soundscape studies': in other words, in this volume there is a two-way traffic between the ethnographic *writing* of sound and the soundscapes of *doing* sound. The methodologies that inform this research are described in greater detail in the following chapter, but while each contribution has a different emphasis one unifying factor is a close practical engagement between the researchers and the soundworlds they study. Thus, we might say that both the writing and the doing of sound take place within an overarching commitment to listening to and with artist-practitioners and audiences. Our aim is to put forward, in Voegelin's evocative phrase, 'a plural practice of touching and singing: the practice of writing as a making of gestures that count the contact between theory, work, and experience in an indivisible sphere' (2021: 280).

1

Sound methodologies in conflict transformation and peacebuilding

Fiona Magowan, Julie M. Norman, Ariana Phillips-Hutton, Stefanie Lehner and Pedro Rebelo

Chapter 1 begins by analysing some of the debates around the challenges of interdisciplinarity and by scoping a range of research methods pertinent to settings that have experienced conflict. We outline how we developed a suite of complementary intertwined methods with our partners in the Middle East, Northern Ireland and Brazil. Each set of methods has been adopted for their critical facility in revealing how the affective properties of sound, music and storytelling can be efficacious 'resources' for expanding 'imagination, awareness, consciousness and action' (De Nora 2000: 24). We elaborate the collaborative and deliberative processes of our participatory action research design with our partners in seeking to reflect the role of sound, music and storytelling in transforming conflict and further enhancing principles of nonviolence (see, for example, Reich 2012). This approach is based on a common set of research questions across the regions, thereby aligning our academic interests with our partners' practice-oriented activities.

The methodology for all the research carried out for the case studies in this book was interdisciplinary, collaborative and complementary. It was interdisciplinary as it combined arts-based practices such as theatre, sound and storytelling, with research from across humanities and social science disciplines, including music, drama, ethnomusicology, anthropology, sociology and politics. We also prioritized dialogue with partners from the outset of the research design to ensure that they were co-collaborators in the development of the methods used. We discuss the process of generating a co-designed suite of distinct creative research methodologies and then explore the integration of different elements and stages of their formation culminating in an inclusive sound art installation. 'Sounding Conflict: A Performance in Five Acts' (see Chapter 6) evokes the themes of resistance, resilience and reconciliation and is co-created with actors from one of our theatre partners who perform its theatrical construction, destruction and rebuilding.

We further compare and analyse the effectiveness of our methods with those discussed in other literatures on interdisciplinary research methodologies (see Chilton and Leavy 2014) and the arts in socially engaged research that seek to produce new and diverse forms of knowledge (see van der Vaart et al. 2018; Wang et al. 2017). In conclusion, this chapter critically assesses the complex dynamics, tensions and ethics posed by our different methods, as well as our approach to integrated multimodal outcomes, as we seek to convey the interrelated challenges and benefits of interdisciplinary socially engaged arts research across protracted and post-conflict contexts.

Listening as methodology and method

> The act of listening is in fact an act of composing. (John Cage)

A move from thinking about sound or music as object to a relational understanding (Born 2010a) that places the sonic event as an agent in a complex network has made listening a key concept in addressing aural experience. Cage's statement captures, in his characteristically succinct manner, the agency inherent in the act of listening. Listening not as a percept of something that is outside oneself but as an active mechanism of sense making, organising, understanding, patterning, and letting loose – that is, composing. Although by no means the only starting point, Cage's work is pivotal for thinking about listening and arguably the springboard for much of the literature and creative practice that most acutely addresses the activity and surrounding discourse since the second half of the twentieth century. Alongside Cage's listening as composing, R. Murray Schafer's ([1977] 1994) concept of 'acoustic ecology' positions listening as a methodology for understanding aural place through the notion of soundscape (Schafer [1977] 1994). The study of the rhythms of everyday life, the listening out for acoustic thresholds, the recognition of *earcons* and varying degrees of transparency in our soundworlds, from lo-fi to hi-fi rely on the act of listening as a method. For Schafer, it is listening that begins the process of framing a region of study, a soundscape or indeed a compositional act. Through acoustic ecology, listening becomes political and environmental, an experiential mode of accessing the equilibrium – or indeed disequilibrium – of a given (sonic) environment. Centring listening within our methodology allows us to consider differences in sensorial relationships and experiential modes that reveal the ways in which listening practices are central to our strategies of research.[1]

One area where this is most evident is our commitment to listening as method. Conducting research on the sonic participatory arts and conflict raises multiple complex ethical issues, notably those around working with victims of violence, forced migrants and children. In addition to the formal issues around data protection this entails, this demands research techniques and methods that uphold Elena Fiddian-Qasmiyeh's (2020) exhortation to avoid reinforcing the political and intellectual power structures that constrain these communities and their representation in scholarship. As researchers in the sonic arts, we were concerned not only with *which* methods would be most appropriate to adopt but *how* we could capture the efficacy of sound, feeling and response. Listening as a practice is thus a core component of how we collect our data in this project, from storytelling to interviews to (despite its ocular-centric name) participant observation. Listening is thus at once methodology and method. As a methodology, listening encompasses a series of ideas about how aural experiences shape perception, understanding and relationship, as well as a way of discussing agency, embodiment and tactility, as the following authors have discussed.

Listening, perception and experience

The Western academic discourse on listening seems to be coated with efforts to disassociate the act of listening from prevailing analogies to seeing – a drive more characteristic of Western philosophy than of the realities one encounters in music practices across the world or indeed in thinking about sound in non-Western cultures. Don Ihde, in his *Listening and Voice*, begins by highlighting how a 'turn to the auditory dimension' represents more than a shift of focus or variables from visual predominance, but rather a 'recovery of the richness of primary experience' (2007: 14). Despite his subtitle 'Phenomenologies of Sound',

---

[1] We recognize the critique of a genealogy of sound studies beginning with figures like Cage and Schafer in particular when it comes to their own political views and standpoints. Their creative practice, however, remains a core reference in thinking through and opening up through sound. With listening practices at the centre of much of what is discussed in the volume, Cage and Schafer stand as key reference points, but they are not the only ones.

Ihde is quick to point out the problematics of the notion of a separate auditory dimension in the first place, given the phenomenological gains in attending to the re-evaluation of the senses as allowing for a recovery of the richness of experience (2007: 21).

The struggle for sensory differentiation is evident even in descriptions of multi-sensorial experience. In his book *Audio-Vision: Sound on Screen*, Michel Chion (1994) draws on Pierre Schaeffer's work on acousmatic music since the 1940s to develop three archetypal listening modes which he distils as 'causal', 'semantic' and 'reduced'. These modes go some way towards articulating a continuum of listening which grows in abstraction. The causal mode is surrounded with a preoccupation for sound and its cause: the 'what's that noise in the middle of the night?' listening. Visualized or acousmatic conditions imbue this mode with either banality (dog is seen, dog is heard) or curiosity (low cry call and response heard in the forest, is it a deer?). As the next point on this spectrum, semantic listening is concerned with meaning and language (be it verbal or otherwise) and is perhaps the mode of listening most comfortably situated within a systems communication model based on a linear flow between transmitter, message and receiver. Finally, reduced listening is the legacy of Schaeffer and the discipline afforded by recorded sound and respective studio-based manipulation – to listen to sound for its own qualities, to appreciate its shape, flow, internal dynamics, texture, timbre and gesture. Reduced listening is perhaps best exemplified when listening derives from a repetitive gesture in which Schaeffer's 'sound object' is mechanically repeated until any concern for its cause or indeed meaning is rendered redundant. At this stage, the listener is left with the arguably infinite richness of sound itself, with listening becoming not reduced in its *nature* but rather focused on sound alone. Chion argues that our listening is framed by the inevitable coupling between sonic event and its cause: the barking dog, the passing car. He juxtaposes the acousmatic versus the visualized to unfold the force of recognizability of source, but this relationship of causality has come under question, including in the traditions of electronic and electroacoustic music since Schaeffer's intervention in the mid-twentieth century. Arguably deliberately a-political, a reduced listening stands as an extreme practice which is set in contrast to much of the *listening through* discussed later in the volume.

Chion's multiple modes of listening have exerted significant influence on the understanding of the relationship between sound and vision, but another tradition suggests that listening to sound is also always a kind of listening to space. Sound needs space to move; physically and acoustically it requires room to propagate, reflect and resonate. As such, a sound is always the space in which it occurs, even when it is a recorded sound played in a space different from where it was recorded. In Steven Connor's elegant formulation, we always seem to be in the space where we listen, what he calls 'ear room' (2010). The relationships between sound and space are rich and complex, and it is no surprise that it has become one of the main arteries of sound practices and studies of the last thirty or so years. The ability for sound to make place, to embed and to change space has fuelled listening practices from Max Neuhaus's LISTEN (1966) to Christina Kubisch's Electrical Walks (2004–). While Neuhaus simply uses the word 'LISTEN' as a way of framing time and space during a walk in which the focus is on the aural experience of the external world, Kubisch's pieces mediate a walk with electromagnetic induction devices, making the listening reach out to sounds not heard by the 'naked ear'.

As these pieces suggest, sonic experience has a shifting relationship with the mechanics of hearing. Seth Kim-Cohen (2009) has even theorized a non-cochlear sound art inspired, in part, by Marcel Duchamp's (1916) critique of painting as merely visual and his proposed non-retinal approach to visual art. Duchamp's *With a Hidden Noise* could even be said to mark the beginning of the non-cochlear sonic art, if there is indeed such a thing. This invitation to imagine sound unheard is presented in an object consisting of two brass plates with screws and a ball of twine containing a secret sonic object. As Duchamp explained in 1956:

> Before I finished it Arensberg put something inside the ball of twine, and never told me what it was, and I didn't want to know. It was a sort of secret between us, and it makes noise, so we called this a Ready-made with a hidden noise. Listen to it. I don't know; I will never know whether it is a diamond or a coin. (Peterson and Sanouillet 1989: 135)

Like Duchamp's imagination of sound, Kim-Cohen's wish to liberate sound from the heard inevitably raises questions on listening. What is there, beyond the aural, that inflects our listening? The politics, the histories, the geographies of culture, everyday life, personal experience and memories that are entangled in our acts of listening shape our experience and our relation to sound, even if unheard.

Discussions of listening – this one included – tend to fall into a duality between the intentional action (listening) versus the passive causality (hearing), yet if we shift our focus from these flavours of individual or social experience to a broader sense of the sonic experience we begin to note permeability and contradiction in what at first seemed the clear-cut binary opposition of listening versus hearing. In the same way that Cage experienced the impossibility of silence in his visit to an anechoic chamber where no sound sources or reflections only served to reveal the sound of one's own body, the act of listening is never laboratory or pure. Rather, it is inevitably mediated by space, time, contexts, histories, technologies and bodies. The simplistic model outlining a direct line from sonic production to hearing and some kind of cognitive process fails to capture individual and contextual conditions. This highlights the multiple ways in which listening is always mediated. As Georgina Born writes, addressing the specifics of music, 'What is it to listen? I have suggested that musical experience (or listening) both results from and engenders mediation' (Born 2010b: 87). Born shifts the focus on listening as something the 'I' does to something that is a condition and activity of a network, a musical assemblage that includes sounds and humans alongside other subjects, discourses, spaces and socialities. By focusing on the relational aspects of listening (English 2017), we reveal how meaning is generated through the web of relationships that flow from a particular object or encounter in both temporal (Schutz 1964) and spatial dimensions (Bourriaud 2002). Pursuing this inquiry, the notion of music or sound as an object to be heard or listened to disintegrates and gives space to an emergent and ever-changing relational experience.

Just as Neuhaus and Kubisch call attention to both the content and mode of listening practices in part by separating out listening from the neurophysiological phenomenon of hearing, so too Pauline Oliveros's practice of Deep Listening reveals the nuances of intent in listening. In doing so, Oliveros extends her own listening to an understanding and care for the listening of others in shared group experiences. In this practice, listening operates in the space-time continuum as sound percepts are interpreted against our memory and mechanisms of prediction. This mindful listening to all sounds opens a space of awareness involving the whole body and mind. Even for non-practitioners of Deep Listening, the recognition of listening as an activity which generates knowledge and builds up a bank of experiences is an important realization. As listening builds on and interacts with previous experience, it can be seen as an interplay between codified sound sets with agreed meaning possibilities (language) and highly personal and unique sonic memory beings that we all are. Likewise, the Buddhist monk Thich Nhat Hanh (2015) uses the term 'deep listening' when he speaks of compassionate listening, imbuing in the act of listening not only empathy but an ability for healing through the very act of sharing. The full attention given to another in a listening session and the transformational effect it can have is evidence that listening is far from a passive act in which a singular message is transmitted to a receiver. As Nhat Hanh writes, 'Instead of "striking" the bell, "invite" the bell to sound' (Thich 2015: 102). Although beyond the scope of this text, the broader field of sound theology is filled with references to how listening practices, music and silence play a role in contemplative traditions across numerous religions (e.g. Foley 2020).

As Oliveros's and Nhat Hanh's practices suggest, listening is an active and ethically charged engagement with the world around us. In Jean-Luc Nancy's unfolding of the language of listening, we discover the layers of significance within such phrases as 'to be all ears' – often condescending or paying lip service; 'to listen in' – to eavesdrop, listening in secret; or 'to be tuned in' – a radio/espionage metaphor for being alert to or in search of. In particular, as Nancy writes, 'To be listening is to enter into tension and to be on the lookout for a relation to self' (2007: 12). As he goes on to clarify, this is the self not as in 'me' but rather the relationship in self, not a metaphor for accessing the self but 'the reality of this access'. Listening thus becomes a listening *to* which garners real affective force.

In *Sonic Agency*, Brandon LaBelle intensifies this sense of listening to by drawing us towards the sonic capacity for community. Here, listening equates with the ability to touch, to reach out, to join and to connect with one another. The vibrational reality of sound is the starting point for a type of listening that is tactile and social before it becomes cognitive. Even more crucially, it is the socially relational nature of listening to sound that grants it transformative power.

> By drawing from experiences and conceptualizations of sound and listening as being conducive to empathy and compassion, as well as the means to break the borders of particular regimes of violence with its interruptive potential, might sonic agency enable an intensification of emancipatory practices? A set of capacities derived from sound's inherently relational force and therefore enabling of new formations of social solidarity, especially as weapons against a neoliberal logic of privatization? (LaBelle 2018: 4)

More recently, in *Acoustic Justice*, LaBelle (2021) addresses issues of the right to speak and be heard working through notions of recognition and resistance.

The notion of the tactile and embodied is also present in Chris McRae's (2020: 399) 'performative listening', in which 'attending to sound and listening as meaningfully informed by the embodied and cultural position of the listener presents an opportunity for framing and practicing listening as a reflexive and ethical act'. McRae goes on to develop properties of performative listening which bring us, perhaps unsurprisingly, close to Cage's quote on listening as composition. Referring to Nina Eidsheim (2015) and her *Sensing Sound: Singing and Listening as Vibrational Practice*, McRae positions listening not as an act that occurs after sound but rather understood as part of the making of sound: 'Listening, in part, makes music' (McRae 2020: 403).

By centring listening as method and as methodology in this research, we recognize the crucial position of the ear as something that may be both revelatory and deeply implicated in our wider sensorial experience. In the chapters that follow, we consider various instances in which the source of sounds cannot be directly identified and explore how these may evoke an interpretive approach marked by curiosity. Of particular relevance is the notion of mediation – and remediation – of sound through technology. By addressing the non-cochlear, we are embracing the sonic as multimodal and we are understanding listening as an act beyond a body-mind dualism. Throughout this book, we *hear* through many examples of listening which is not about the essence of the sonic experience but an act of engagement in which the sonic acts as a conduit for the political. Our understanding of listening, especially in the context of a methodology (used here to signify the rationale for our research approach), is of a multilayered act of intent. Yes, listening can involve hearing as a sensory activity but it also involves other senses and a cognitive engagement which is framed by the individual, the social, and the relational ties that bind.

Research design

The broader project uses a grounded theory approach, in which theory is 'inductively derived from the study of the phenomenon it represents … One does not begin with a theory, then prove it. Rather, one begins with an area of study and what is relevant to that area is allowed to emerge' (Strauss and Corbin 1990: 23). Within this inductive framework, we adopted a mixed methodology approach, incorporating the case study framework with participant observation and narrative interviews. Greene et al. (1989) identify five purposes for using mixed methods, which are in accordance with the aims of this research. First, triangulation, or seeking convergence of results, reflects our goal of drawing informed insights on arts-based acts of resistance, resilience and reconciliation by comparing results from different sources. Second, complementarity, or examining different aspects of a phenomenon, is evident in this study's emphasis on considering various examples of arts-based resistance, resilience and reconciliation in conflict contexts. Third, initiation, or discovering fresh perspectives, is evident in this project's use of multiple methods to communicate how untold stories generate new ways of conceptualizing arts-based resistance. Fourth, development, or using methods to inform each other, is evident in this study's use of preliminary observation to inform interview questions. Finally, expansion, or adding scope to a project, is reflected in this research by incorporating different methods to expand the study beyond individual narratives to draw conclusions about collective community experiences (Greene et al. 1989; Norman 2010: 81; Tashakkori and Teddlie 1998: 43).

Each section of the project adopted a case study approach. According to Robert K. Yin (2003a: xi), case studies are useful when researchers seek '(a) to define research topics broadly and not narrowly, (b) to cover contextual or complex multivariate conditions and not just isolated variables, and (c) to rely on multiple and not singular sources of evidence'. This research was conducted in accordance with these conditions, as we sought to explore sound-based phenomena in various contexts with diverse realities, relying on different data sources. As Yin summarizes elsewhere, 'Case studies are the preferred method when "how" and "why" questions are being posed, when the investigator has little control over events, and when the focus is on a contemporary phenomenon within some real-life context' (2003b: 1).

For each case study, integrating the specific methods of narrative interviews and participant observation was particularly useful. As Joseph Maxwell notes, 'While interviewing is often an efficient and valid way of understanding someone's perspective, observation can enable you to draw inferences about this perspective that you couldn't obtain by relying exclusively on interview data' (2005: 94). Indeed, observation was important for assessing the extent to which ideas communicated in interviews were put into practice and for revealing perspectives that interviewees might have been reluctant to state in interviews. On the other hand, although observation provides a direct way of learning about behaviour, 'interviewing can also be a valuable way of gaining a description of actions and events' (Maxwell 2005: 94). For example, interviews were useful for learning about the backgrounds of both the individuals and the organisations we engaged with that informed their present actions and activism. In addition, interviews gave insights on events to which we could not gain access, such as private meetings or performances, or actions that took place when travel was restricted. Thus, as Maxwell notes, 'Triangulation of observations and interviews can provide a more complete and accurate account than either could alone' (2005: 94).

Participant observation

Danny L. Jorgensen (2015) describes participant observation as the process by which 'the researcher interacts with people in everyday life while collecting information, [constituting] a unique method for investigating [their] enormously rich, complex, conflictual, problematic, and diverse experiences'. Participant observation was crucial for gaining understanding of the history, culture and context of each of the case studies. For example, in our project in Lebanon, (Chapter 2) this included elements that contributed to the 1975–90 conflict and the 2019–20 revolution. Acting in different contexts as participant observers and observing participants (Tedlock 1991), our immersion in the field site consisted of attending performances, workshops, meetings and dialogue sessions of our two primary community partners, Fighters For Peace (FFP), an NGO comprised of ex-combatants working to prevent violence, and Laban, a community theatre. We also observed protests, demonstrations, concerts and arts installations taking place in Beirut and Tripoli. We documented events when possible with video, sound recording and/or photography, and wrote up observations in daily field notes.

Participant observation is often immersive and deeply engaging; in working with Musicians Without Borders (MWB; Chapter 3), our participant observation included learning a range of vocal sounds and singing to develop leadership skills, rhythmic percussion and entrainment, such as in drum circles, understanding movement motivation, group songwriting and co-creating, and performing in a trash orchestra. We experienced how the techniques employed in the workshops could expand and enhance existing capacities of the trainees, as well as enabling us to reflect upon our own sonic and musical engagement. For example, in one exercise, sheets of paper were distributed around the room, and participants were invited to make sounds with them in every way possible. We began rustling, crinkling, quickly ripping, gradually tearing and crushing paper while others blew on the edges creating high pitched shrill squeaks and yet others rubbed them together to create rhythmic patterns. Sound dynamics were determined by the facilitator who indicated when they should be hushed or more explosive. The effect of this sonic composition was to create enhanced interpersonal intimacy that developed senses of 'curiosity, vulnerability, empathy and the recognition of irreducibility' (Obert 2016: 26). These activities consolidated cooperative rather than competitive relationships as participants learned how to improve their capacity to work together through 'whole body or somatic experiences' shared with one another (Bang 2016: 358). As researchers, we were directly challenged by some of the activities which pushed us and some of the participants out of our comfort zones into new spaces of improvisation (see Chapter 3 for further discussion).

Ultimately, participant observation was important for several purposes. First, the contextual awareness helped us in understanding and interpreting the data we acquired through our interviews and other sources. Further, being present at various actions and events introduced us to arts- and sound-based activism strategies that we would not have encountered or fully understood through interviews alone, such as the dynamics of playback theatre or storytelling sessions. Finally, being participant observers increased our access to activists and artists, as we made contacts simply by being present at events. Our ongoing presence also helped enhance our trust relationships and sense of legitimacy with activists who might have been reluctant to speak with us via a one-off interview.

Narrative interviews

Several parts of our research relied on narrative interviews to gain more in-depth understanding of the arts-based resistance initiatives, and to ground the research in personal experiences with those initiatives. As Jean Clandinin and Michael Connelly (2000) argue, narrative research is uniquely capable of capturing individuals' stories and investigating how they perceive their experiences in the temporal, spatial and personal-social dimensions (see also Norman 2010: 86). Furthermore, when considered collectively, interviews can indicate how individual, group and community stories intersect to inform – and be informed by – social phenomena.

Indeed, narrative research probes beyond the mere reporting of events, and even beyond the individual's role in or opinion of such events to offer multiple dimensions of analysis. Across this project, interviews were semi-structured drawing from a general list of questions and themes, but with space for participants to speak in depth on often difficult subjects without pre-direction. This allowed the researchers to ask follow-up questions or pursue unexpected lines of inquiry as well as helping develop positive rapport. By generating a free-flowing conversational dynamic, these interviews fostered a safer space for revealing emotions, past experiences and controversial opinions (Norman 2010: 95).

The semi-structured interview approach is especially useful for understanding the individual experiences and opinions of diverse interviewees, while also allowing for comparisons and generalizations between participants. Many interviews also included opportunities for participants to ask questions of the interviewers. With permission, interviews were recorded, transcribed and analysed to identify representative quotes and themes. Where appropriate, these transcriptions and the ensuing analysis were then shared with the participants as a way of upholding principles of collaboration and transparency in research. These opportunities for dialogue during and after the interview process reinforced the narrative interview as a collaborative process in which both participants are engaged in learning, and it also encouraged a relational approach to research (informed in part by the writing of Rihab Azar [2020]) that rejects the goal of representation of an 'inside' participant by the 'outside' researcher in favour of a simultaneously humble and compassionate mutual inquiry.

In the project's initial stages, interviews were conducted in person by the authors or in collaboration with community partners. In the latter stages, as fieldwork and travel were limited due to Covid-19 pandemic restrictions, interviews were often conducted over videoconferencing technology such as Skype, Zoom or WhatsApp, and were supplemented by email interactions both before and after the interviews. The combination of video and email correspondence was particularly important due to the difficulties in assessing understanding and non-verbal cues across videoconferencing technology as well as allowing for continued conversations.

(Co-)creative methods

The range of case studies in this volume encouraged a similar breadth of method in which listening to the experiences of others through in-depth narrative interviews and storytelling was paired with other methods such as surveys, group discussions, critical and textual analysis and practice-based reflexive analysis. These traditional methods of data collection and analysis were employed in conversation with what are known as creative methods. Creative methods thrive on an embodied engagement with a given context and often embrace risk, improvisation, iteration and constant re-examination. They also recognize the inevitable shift that occurs upon intervention, ever rejecting the illusion of neutrality. As the primary *modus operandi* of most of our partner organizations, which range from theatre companies and community music institutions to composers and community museums, engaging with creative methods allowed us to more accurately understand and respond to the activities of our partners. Within this methodological breadth, the specific choice of methods was adapted to suit each case study. For example, Chapter 4 focuses on musical compositions in the contemporary Western avant-garde tradition which are a more evidently 'presentational' (Turino 2008) and mediated type of sonic art than those discussed in Chapters 2 and 3. Historically, musicologists have studied so-called Western art music using historiographical and analytical methods that prioritize textual representation (e.g. a written score) and critical reception, but here these methods were brought into conversation with the ethnographic exploration of participant experience. In contrast, Chapter 6 articulates how creative practice can dialogue with other research methods in the context of the creation of the installation 'Sounding Conflict: A Performance in Five Acts'. It also discusses how artistic interventions in participatory contexts can reveal insights often difficult to access through more conventional methods.

In each of the case studies, we refined our methodologies to address specific effects. For example, in the MWB project (Chapter 3), we initially used a 'graphic elicitation method' (Palmer et al. 2014: 527) as 'a participatory method and as a useful tool for thinking' (Heath et al. 2018: 713) in the form of an emotion curve survey that sought to trace the trajectory of ten emotions over the first five-day programme (see Donaghey and Magowan 2021). While this linear graph provided a sense of the intensity of different emotions, it was a blunt tool to capture a relatively long period and acted thus as a general mnemonic of emotions. We further refined this methodology during the second MWB programme to elicit specific emotion responses to a ten-minute 'music and movement improvisation game (a normal part of the training schedule), rather than reflecting on the whole training program' (Donaghey and Magowan 2021: 4). Participants used this exercise as a form of creativity expressing and reflecting upon their emotions in graphic designs accompanied by textual elaboration of how the designs represented these emotions. The purpose of these emotion curves was to create new kinds of 'imaginal methods' that 'take into account and engage emotion and affect in the process of transformative learning' (Dirkx 2006: 24). As this suggests, participating alongside our partners allowed us to be responsive and creative in co-creating methods.

Although each of the artistic practices discussed in this book carries their own lineages and communities of practice, there are cross-cutting threads that enable us to speak of creative methods as contributing to a landscape of increasingly multi-method and multimodal approaches to research. Terms such as 'practice-based research' and 'artistic research' encapsulate a multitude of research processes and behaviours which have a creative output at its core. Across music, theatre, design, dance, film and the visual arts, the process of generating knowledge through creative practice is not free from complexities and limitations. For example, as the core of this type of research resides in the creative outputs themselves, these are inevitably tied to specific discourses surrounding a given artistic community. As such, artistic research is intertwined with other methodologies and art works acting both as starting points and as results in the research process.

As Julian Klein (2017) writes, 'At what level does the reflection of artistic research take place? Generally, at the level of the artistic experience itself.' To engage in research that is articulated through the 'artistic experience itself' shifts the focus of knowledge generation from the written to the sensorial. While this does not prevent textual engagement, analysis, and critique, the experience remains the space of knowledge which is often unique to a given artistic practice and which, by definition, cannot be expressed in other forms. Throughout the book we discuss various sonic expressions of creativity, at times framed within specific art forms, at times through more fluid manifestations of the creative act, on some occasions indistinguishable from the everyday, releasing the 'artistic experience' from an institutionalized drive to frame the concert, the performance or the exhibition. These everyday creative experiences are embedded in acts of listening and acts of making, with storytelling as a common thread in many of the artistic expressions discussed here. The expressions are intrinsically woven into the everyday through creative practice. As a practice, it has no beginning nor end, it is ongoing and ever questioning and framing our relationship to the world.

Challenges and limitations

As with all research designs, there were challenges and limitations inherent in the chosen methods in terms of scope, scalability and objectivity. First, in terms of scope, our focus on participant observation and narrative interviews is more useful for achieving depth of knowledge rather than breadth. This was partially balanced by the incorporation of audience surveys and other methods of analysis, but research with a different scope might highlight alternative features of sonic participatory arts. Moreover, the organizations that we partnered with and the pieces that we studied are not necessarily representative of the wider society, which might be seen as a limitation. However, it is specifically because these organizations and individuals are doing disruptive, non-traditional work that we wanted to engage with to understand their motivations, practices and impacts. Second, and similarly, each case study is rich in insights, but not all of our conclusions will extend to the other cases. Accordingly, there is little suggestion that the individual experiences or compositional practices detailed here should be taken as a template; nonetheless, exploring each of these studies in some depth and juxtaposing them against the other case studies in this book helps us better identify what lessons might be extrapolated and which are contextually specific.

Third, it is important to recognize potential concerns regarding objectivity. As noted above, one of the benefits of collaborative, participatory research is the cultivation of trust relationships that can allow for greater access and more honest exchanges. However, those same relationships can make researchers prone to bias in their interviews, especially when using a non-standardized format, or even in their interpretation of the data. We aimed to minimise and address this effect again through the collaborative nature of the project across the researchers. We ensured there were some standardised questions in the interviews, and we shared and discussed our field notes, drafts and recordings with other team members outside the case study research for their feedback and analysis. Nevertheless, we acknowledge that the research is not purely objective, in that it seeks to develop pragmatic theories related to social change that will be useful to community partners and communities living in or emerging from conflict.

In addition to questions of scope, scale and objectivity is the effect of the Covid-19 pandemic on this project, which stretched beyond the restrictions on the travel necessary to conduct in-person research to encompass minimized rehearsal time for musicians and artists, challenges establishing live-streaming technologies, and the wholesale cancellation of performances. In particular, the cessation of live musical and theatrical performance during this period meant that there were limited opportunities for the researchers to follow up on theatrical performances in Derry and Belfast (Chapter 5) or to put on planned performances of 'Sounding Conflict: A Performance in Five Acts' in Brazil (Chapter 6). In the case of Chapter 4, the pandemic also forced a change in the research subject: initially the plan was to examine one of Mary Kouyoumdjian's large-scale compositions, but that premiere has been delayed, and the decision was taken to focus instead on an earlier, smaller scale piece that could be accessed via audio and video recordings.

The lack of co-presence and the alteration of in-person experience occasioned by the pandemic continues to resonate strongly among all of the sound-based creative arts and those who participate in them. Yet this may heighten our awareness of the many ways of sensorially engaging with one another. For example, while there are inevitably limitations to studying behaviour and motivation via creative activities, participants in these cases offered rich reflections that generated 'deep insight by going beyond rational-cognitive ways of knowing' (van der Vaart et al. 2018: 1). By engaging multiple senses, participants were channeled into processes of deep listening and responsive, and directed interplay with one another that evoked 'emotional visceral response[s]' (4). The combination of multiple creative methodologies in this project enabled a wider approach to more standard research methods facilitating a complex layering of meanings to emerge in the process. Our creative methodologies also enabled us to work appropriately through the lens of participatory action research by working together *with* participants and thereby giving voice to their experiences in the moment rather than employing post-activity memory capture and recall of the effects of the creative practice retrospectively.

Conclusion

The narratives traced throughout this book are a manifestation of our broad interest in storytelling as a sounded-out means of (de)stabilizing identity, generating interpersonal relationships and co-creating meaning in post-conflict societies. This approach is crystallized within the work of the philosopher Adriana Cavarero (2000), who points out the impossibility of autobiography in her exploration of the importance of intersubjective relationality. Given that no one can tell the tale of their own birth (much less their own death), each individual is reliant on others to tell their story. As Cavarero notes, the desire to listen to one's story points to the fundamentally dialogic character of narration and the necessity of relationship in sustaining the stories that are told about individuals and about the world. Likewise, as Paul Ricoeur explains, narratives are both lived and told, mediating between the world of action and the world of recollection/interpretation (1984; see also Norman 2010: 87). The range of ethnographic and creative methods we employ to collect, analyse and retell the stories of our research partners complements the focus of the entire project on sound-based expression in conflict, as the research itself engages with a praxis of telling, performing and listening.

As Victoria Foster (2012) has noted, these kinds of creative arts-based approaches are often considered to be less tangible than other social science methodologies. Yet the latter also tend to overlook the importance of embodiment in understanding the impact of practices upon participants. Thus, impact is frequently conceived of in limited terms rather than adopting more expansive perceptions of the effects of co-produced arts approaches to sensory ways of knowing that can also be more democratising in their approach. Yet, as we suggest above, these creative arts-based approaches play into a more expansive and critical methodology of listening. The unfolding of this methodology through the following chapters promises a rich panoply of analytical entry points into sound and conflict that embraces all kinds of participation in the sonic arts as moments of 'encounter' (Cook 2012). Taken together, they suggest a multitude of ways in which the sonic arts push us to think creatively about resistance, resilience, remediation, reconciliation – and beyond.

2

Resistance: Performing the frontline

Julie M. Norman

Stories are very powerful and when they come to life, they are important and make people feel that they matter, and they deserve to be heard, and that will make them get louder and make sure someone is hearing.
    –Farah Wardani, theatre director and activist, Lebanon.

Introduction

How do narrative, theatre and storytelling inform resistance during and after conflict? How do various forms of sounding conflict both reflect and influence social and political change? In this chapter, we explore the concept of resistance in different ways, from political protests, to protracted transitional justice struggles to everyday acts of steadfastness. We examine how narrative, theatre and storytelling, as well as music and sound, function to mobilize activists, amplify causes and grievances, and bring communities together. At the same time, we discuss some of the challenges and limitations to sound-based interventions in realizing societal change in some contexts.

The chapter is organized as follows: first, we examine the meaning of resistance and the ways in which resistance intersects with 'sounding conflict', focusing on the *relational*, *everyday* and *spectral* elements of resistance. Second, we situate the concept of sounding resistance in the context of one of our core case studies – Lebanon – recognizing both the strengths and weaknesses of sound-based approaches in various manifestations of resistance. Finally, we discuss how sounding resistance interacts in integral but complicated ways with processes of resilience and reconciliation in protracted conflicts and post-conflict environments.

Concepts of resistance

There is no single definition of 'resistance' in scholarship or in practice, and the term has varied meanings for different individuals and in different contexts. In academic literature, the concept of resistance is inherently interdisciplinary, drawing insights from sociology, political science, psychology, anthropology, geography and law, and applied to individual confrontations, protests, revolutions, social movements, class struggles and more. In practice, resistance, like activism, is manifest in myriad actions, tactics, strategies, movements and performances. While we acknowledge that it is neither possible nor desirable to hone in on just one definition, for the purpose of examining sound and resistance, specifically, we find it useful here to unpack the *relational* nature of resistance, especially with regard to power and also the concept of *everyday* resistance. These aspects cut across our case studies and are integral to the ways in which storytelling, music, theatre and sound relate to various manifestations of resistance in conflict.

Relational nature of resistance

The relational nature of resistance, specifically the relationship between power and resistance, was articulated by Michel Foucault (1978: 95), who famously noted, 'Where there is power, there is resistance, and yet, or rather consequently, this resistance is never in a position of exteriority in relation to power'. However, as Mikael Baaz et al. (2016) rightly observe, Foucault primarily emphasized the power element of the power/resistance relationship in his analysis, rather than the resistance element, thus requiring further interrogation as to how resistance manifests in relation to power. Gene Sharp (1973: 4), writing on nonviolence, also situates resistance in relation to power, noting that 'the exercise of power depends on the consent of the ruled, who by withdrawing that consent, can control and even destroy the power of their opponent'. In other words, power is fluid, and when people resist, even if they are repressed, they challenge the status quo power dynamic because of its relational nature.

More recently, Jocelyn Hollander and Rachel Einwohner (2004) also emphasize the relational, or 'interactional' nature of resistance. In their thorough overview of resistance literature, they conclude that, across definitions and conceptualizations, resistance includes a sense of *action* and a sense of *opposition*. Crucially, they continue, these aspects raise questions about *intent* and *recognition*, stating, 'Resistance is defined not only by resisters' perceptions of their own behaviour, but also by targets' and others' recognition and reaction to this behaviour … understanding the interaction between resisters, targets, and third parties is thus at the heart of understanding resistance' (Hollander and Einwohner 2004: 548). They offer a helpful typology of three groups, namely actors, targets and observers; further, they discuss how actions might be classified as different forms of resistance depending on if the act is *intended* as resistance and if it is *recognized* as resistance by targets and/or observers. For example, 'overt resistance' occurs when the act, such as a public protest, is intended as resistance by the actor and recognized as resistance by both targets and observers. In contrast, 'covert resistance' is when the act, such as foot-dragging or go-slow worker movements, is intended as resistance and recognized as such by observers but not recognized by the target. Drawing from Foucault, they note that analysing the relational nature of resistance in this way 'highlights the central role of power, which is itself an interactional relationship' (Holland and Einwohner 2004: 548).

The typology developed by Hollander and Einwohner is especially helpful in the context of sound and arts-based resistance; for example, if a sound or voice is not heard, is it still resistance? And what is the role of a potential audience (or 'observer') as a conduit between actors and targets, especially in theatre, storytelling and performative spaces? Tamra Pearson d'Estree's (2005) analysis of the role of voice in conflict and conflict resolution complements Hollander and Einwohner's typology by articulating four levels of 'voice'. According to d'Estree (2005: 105), voice can be understood from the psychological perspective as the 'expression of one's thoughts and opinions', or from the social narrative perspective as 'the ability to tell one's story'. Further, d'Estree (106) notes that voice is inherently intertwined with 'status, identity and even existence', such that 'to have voice is to have legitimacy and integrity; denial of voice denies … legitimate participation, ruptures integrity, and threatens the self's existence'. In conflict contexts then, d'Estree identifies Level One of voice as 'the ability simply to voice one's views and experiences' (110). However, just as Hollander and Einwohner question the impact of resistance if it is not recognized by targets or observers, d'Estree rightly asks, 'Is it merely enough "to voice," or … is it the social implications of having a voice and being listened to that actually matter?' (110).

The subsequent levels of voice identified by d'Estree thus take a more relational approach, noting that, like the validation of resistance, the 'validation of voice may come through hearing by an interested outsider, by members of the perpetrator's group, or by the community or society' (2005: 111). Level Two focuses on the 'ability to be heard by others' (111), aligning with Hollander and Einwohner's 'observers', while Levels Three and Four emphasize the ability to be heard (Level Three) and acknowledged (Level Four) by the other, who perpetrated the injury (112), similar to Hollander and Einwohner's 'targets'. Integrating these theories of resistance and voice with sound can be represented as follows:

Resistance		Voice/Sound
Actor/agent/activist	←→	Narrator/performer
Target	←→	Perpetrator/oppressor
Observer	←→	Community/audience

As our case studies indicate, these categories are not fixed or mutually exclusive but rather fluid and overlapping. However, this typology can help us integrate the relational concepts of resistance, voice and sound to better understand how these processes intersect in theory and practice.

Everyday resistance

The term 'resistance', especially in a conflict context, may bring to mind images of protests, rebellions and revolutions. However, in each of the contexts included in this project, acts of resistance are also manifested in everyday forms, especially with regard to sound and arts-based resistance. As one of the seminal scholars on everyday resistance, James Scott (1985: xv) states, 'Where institutionalized politics is formal, overt, concerned with systematic, de jure change, everyday resistance is informal, often covert, and concerned largely with immediate, de facto gains'. As Scott continues, in contrast to more visible social movements, these forms of struggle 'require little or no coordination or planning; they make use of implicit understandings and informal networks; they often represent a form of individual self-help; they typically avoid any direct, symbolic confrontation with authority' (xvi). It is helpful, however, to extend beyond Scott's conceptualization of everyday resistance, which he defines as mostly individual, uncoordinated and covert, to recognize that such tactics often complement or shift into open dissent or confrontation (Adnan 2007; Norman 2020). This understanding of everyday resistance, as gradual and unspectacular, but still coordinated, and, at times, confrontational, is especially useful when considering the role of sound, story, theatre and music in resistance, in which actions (or performances) may be relatively subtle yet still challenge authorities.

Indeed, Johansson and Vinthagen (2020; Johansson 2016) offer a more recent and applicable discussion of everyday resistance, basing their analysis on the assumption that everyday resistance is a practice; it is historically and intersectionally entangled with power; and it is variable in different contexts (Johansson and Vinthagen 2016: 418). They thus propose a framework based on repertoires of everyday resistance, relationships of agents, spatialization, and temporalization of everyday resistance (419). We extend this framework by situating it within the context of sonic resistance, noting the specific repertoires, power relationships, and spatial and temporal implications of resistance rooted in sound, music, storytelling and theatre, especially in conflict settings.

Indeed, the concept of repertoires is especially useful when exploring the role of sound in resistance. Charles Tilly describes 'repertoires of contention' (McAdam, Tarrow and Tilly 2001; Tilly 2004) as the set of tools or actions available to a movement in a given context; in our case studies, this includes sound-based approaches. While repertoires emerge in relation to the opportunities or constraints imposed by the given context, we adopt Johansson and Vinthagen's view that activists, including artists, musicians and storytellers, employ innovations (2020: 105) that are often creative and not just reactive (Norman 2020). Indeed, the concept of repertoire allows us to extend the notion of resistance tactics to arts-based approaches that might be overlooked in traditional conceptualizations of resistance.

The other aspects of Johansson and Vinthagen's framework for everyday resistance are also helpful for understanding sound and resistance. Like Hollander and Einwohner, they emphasize the relationships of agents and actors, which, as we noted, is especially nuanced when considering spaces of sound, story and/or performance. Further, the notions of spatial and temporal implications of resistance are also especially useful when considering sonic resistance. In terms of temporality, some forms of sound-based resistance have exceptional poignancy or strength because they only exist in the moment in which they are sounded or heard, while other forms, such as those that are recorded or passed down through oral tradition, can transcend temporal boundaries that might constrain or limit other forms of activism. Spatially, sound can penetrate spaces that are physically separate or off-limits, while there are also limitations to the extent to which sound can be present in other spaces.[1] This framework of everyday resistance underpins the analysis of sound-based resistance highlighting some of its strengths and limitations.

Conceptualizations of resistance on spectrums

In both theory and practice, the notion of resistance is not absolute but rather entails 'dimensions of variation' (Hollander and Einwohner 2004: 537) or 'spectrums' (Baaz et al. 2016: 146) of resistance. Such spectrums are especially useful when considering diverse case studies that include different approaches to resistance in different places and time periods. Spectral considerations are also helpful when analysing sound and arts-based expressions of resistance, which may be less overt or identifiable than typical resistance tactics but which still contribute crucially to processes of social and political change. They can be identified according to three broad parameters:

Individual–Collective

As Baaz et al. (2016) note, the spectrum between individual and organized resistance extends from individual actions such as whistle-blowing to majority mobilizations such as anti-colonial struggles, as well as to intentional movements coordinated by civil society organizations or transnational advocacy networks. In considering arts and sound-based resistance, the individual-collective spectrum can be quite fluid, starting with individual artists, musicians and performers who may become part of a collective in the course of a performance and whose performance may inform an even broader collective in terms of the audience and recipients it reaches. The same is also true with storytelling and narrative, as the individual's story informs the community narrative. As Marshall Ganz (2007) explains, sharing personal narratives, or 'stories of self', allows for the development of 'stories of us' within a community, finally leading to a 'story of now', or plan of action based on personal experiences and community values. The individual-collective spectrum is also reflected in d'Estree's levels of voice, as discussed above,

---

[1] Sound was key to overcoming spatial boundaries during the Covid-19 outbreak, during imposed lockdowns and physical distancing. In communities where people could not physically meet or see each other, sound and music were frequently used to show social solidarity.

which extend from individual expression to being heard by others, indicating how certain voicings and soundings can be both individual and collective.

Covert–Overt

Related to the individual-collective spectrum, resistance can also be conceptualized on a spectrum ranging from the discreet to the spectacular. Along this continuum, everyday resistance can range from the covert, as articulated by Scott (1985), to the more sustained and disruptive. Thinking of resistance in this way is especially crucial for arts-based practices in general and sound-based expressions in particular. For arts-oriented resistance, some mediums lend themselves to being more public and visible than others, and some artists/performers seek to make their art more intentionally provocative than others. Indeed, for some artists/activists, the fact that the resistance is discreetly 'hidden' within the art form may be one its strengths, so that it can avoid detection from authorities or opponents, as further articulated by Scott's work on hidden transcripts (1990). For example, during the Pinochet dictatorship in Chile, activists and artists hid political messages and maps in safe houses in *arpilleras*, traditional sewn handicrafts, evading notice from the regime because of the seeming simplicity of the medium.

With sound-based expression, the absence of the visual component requires activists and artists to consider the literal volume of their resistance or protest. Will they embrace a whispering campaign to communicate covert messages and mobilize support, or will they bang pots and pans from balconies and rooftops as a gesture of defiance or solidarity? Will they share narratives through small storytelling circles or through performances in public spaces? The covert-overt spectrum of sound-based resistance also has a temporal dimension. While visual art typically has some given duration, whether in physical and/or digital form, sound-based resistance can be limited to its moment of expression. At the same time, if recorded, that moment can echo and extend across time and space, such that a performance that was initially covert can reverberate to become more central and overt.

Process (expression)–Outcome (impact)

Another spectral consideration is the extent of emphasis on the act or process of resistance in contrast to its impact or outcome. That is, is activism still 'worth it' if it fails to realize all or some of its objectives? Is mobilization and expression an end in itself, or must a movement achieve tangible goals to be considered successful? This spectrum of emphasis is especially notable in arts-based forms of resistance. As Julie M. Norman (2009) notes, participatory media, a form of arts-based resistance, can achieve real activism and advocacy goals by challenging dominant discourses, offering alternative information sources and creating spaces for critical dialogue. However, tracing and measuring such impact can be difficult, especially if the medium itself is intangible, as with sound-based media. Nevertheless, as Norman explains, such arts-based resistance can also foster opportunities for creative expression and civic engagement, especially in spaces in which participation or activism is restricted or marginalized. Similarly, as Stephen Duncombe (2007: 490) writes, 'Resistance expressed culturally can engender solidarity, create a shared set of norms and values, and be the jumping off point for imagining new communities and political subjectivities'. This 'possibility of movement *from* cultural resistance *to* community development' (490) suggests that arts-based activism can be a means to an end but an end that is more rooted in building community and solidarity than in achieving a specific policy objective, though that may sometimes be the case as well.

A related spectral consideration is the relationship between vulnerability and agency in resistance. Judith Butler (2016: 13) rightly asks, 'Does resistance require overcoming vulnerability? Or do we mobilize our vulnerability?', and 'Does the discourse of vulnerability discount the political agency of the subjugated?' (22). Butler concludes that 'vulnerability can be a way of being exposed and agentic at the same time' (24); that is, individuals or groups may be in objectively precarious situations, especially in conflict contexts, but this does not fix them in a state of powerlessness. Rather, vulnerability can inform social and political activism, such that 'vulnerability enters into agency ... and the binary opposition between them can become undone' (25). Breaking down this binary is especially crucial in understanding arts-based and sound-based activism, in which performance spaces allow for the expression of vulnerabilities in a way that exerts agency. As Charles Tripp (2013: 204) notes, 'Performance [is] not extraneous but rather implicated in the very notion of political behaviour itself as a social act'.

Resistance – resilience – reconciliation

The final spectrum to consider is the one that runs through this book – that is, the relationship between resistance, resilience and reconciliation. To what extent is resilience a form of resistance? In contrast, as Emma Keelan and Brendan Browne (2020) ask, when does advocating for resilience stifle resistance or activism? With regard to reconciliation, can acts of resistance also facilitate reconciliation, or are the two diametrically opposed? In this chapter, while the focus is on resistance, the actions observed and discussed from the case studies are not distinct from processes of resilience or reconciliation, but rather interwoven with them in sometimes challenging ways. The resistance lens is crucial to understanding how sound, music, theatre and storytelling manifest during and after conflict, but our case studies indicate that the resistance component does not operate in isolation.

Lebanon case study: From confronting the past to shaping the future

The spectral, relational and everyday elements of sound and resistance were evident in our various case studies, revealing how sound-based practices manifest differently across diverse conflict contexts. We have observed the use of sound in relation to resistance across the spectrums of individual-collective, overt-covert and process-outcome–oriented elements, as well as, crucially, in blurring the lines between resistance, resilience and reconciliation. Further, many of our case studies reflected the relational nature of resistance, with actors/narrators/performers using sound to express their voice to targets/perpetrators/oppressors often against the backdrop of observation by wider audiences/communities. In addition, multiple cases reflected elements of everyday resistance, with activists/artists/musicians using a repertoire of tactics to challenge spatial and temporal boundaries. In this section, we turn to the case study of Lebanon to illustrate how sound, music and narrative were integral to resistance and where resistance itself was intertwined with resilience and reconciliation.

Background

Our research in Lebanon focused on the activities of two complementary organizations, the NGO, Fighters for Peace (FFP), and the improvisational theatre group, Laban Lactic Culture. When significant street battles erupted in Tripoli, Lebanon, in 2014, five former participants of the Lebanese civil war (1975–89) from antagonizing militias issued a joint statement to urge those fighting to stop. The public response to this statement was overwhelmingly positive, and as a result the group formed FFP, an organization comprised of ex-combatants committed to preventing political violence. FFP includes fighters from different political, religious and social backgrounds and has undergone significant introspective processes that resulted in renouncing violence and confronting the trauma resulting from their participation in conflict. Through their multi-perspective narratives on war and personal change, the ex-combatants act as credible, authentic role models who can connect to former fighters or at-risk recruits, as well as to society in general, to promote dialogue, reconciliation and nonviolence.

FFP's activities focus on individual and collective storytelling to confront the past and, through sharing their experiences, prevent violence in the future. Based on my observations and FFP's own description of their work, they engage in a range of narrative-driven outreach activities:

- Dialogue sessions held in schools, universities and community centres: Ex-combatants give their testimonies, highlighting the changes they have undergone and lessons learned from violence, and engage with the audience in open and frank two-way exchanges.
- Individual exit assistance and rehabilitation for current and recent combatants, in which FFP offers expert mentorship to help 'formers' (ex-fighters) deal with trauma and reintegrate into society.
- Insider mediation in divided communities and at-risk locations with key members of civil society, community leaders and former combatants.
- Transitional justice processes: Cooperation networks are created with groups representing victims of the war such as the Committee of the Families of the Disappeared, ACT for the Disappeared and International Committee of the Red Cross (ICRC), who action transitional justice collectively.
- An online archive of oral history interviews charting Lebanon's civil war to contribute to collective memory and offer alternative narratives in the absence of national reconciliation.

In this research, we focused on FFP's use of participatory storytelling via engagement with Laban, a community theatre company based in Beirut. Laban launched a Playback theatre programme in 2013 and began working with FFP in 2016. Playback is a form of improvisational theatre in which a person shares their story, experience or memory, then actors recreate the story artistically to give it shape.² A facilitator or 'conductor' then facilitates a discussion with the storyteller, the actors and the wider audience. Developed in New York in 1975, Playback has been used internationally in educational, therapeutic and community spaces. FFP's collaboration with Laban started with fighters sharing their stories from the civil war, but it has grown to include the stories of victims and survivors from the war. They have also collaborated to tell stories from contemporary conflicts such as the Syrian civil war.

Neither FFP nor Laban specifically uses the language of resistance. In Lebanon, the term 'resistance' (*muqāwamah*) typically refers to Hezbollah's armed resistance against Israel, or, alternatively, to personal or psychological resistance to interventions or healing. Activists instead use terms such as 'activism', 'social change', or in the 2019–20 period, 'revolution' (*thawra*). In reality, FFP and Laban have always been engaged in a sort of activism in terms of confronting the state-level amnesia to the civil war and creating spaces for reconciliation, as discussed further on in this chapter. In the next section, however, we discuss the integral role of sound, storytelling and theatre in the popular uprising of 2019–20.

Context: The 2019 uprising

Amid a worsening economic situation, protests broke out in Lebanon on 17 October 2019 when the government announced new tax measures, including a new tax on internet-based voice-call services like WhatsApp. Although the tax was later withdrawn, demonstrations increased, spreading from Beirut throughout the country. Demands included ending government corruption, ending the sectarian political system, recovering stolen funds and holding corrupt politicians accountable and establishing fair tax and financial procedures (Amnesty International 2019). Prime Minister Saad Hariri tendered his resignation on 29 October 2019, but protesters, pushing for systemic change, continued protesting into the early months of 2020. The protests were notable not only for their numbers, with hundreds of thousands demonstrating, but also for their unified nature in a usually divided Lebanon, with participation from nearly all sects, religions, geographic areas, social classes, genders and age groups.

---

² 'What is Playback Theatre?' Playback Theatre Southwest. Available online: http://www.playbacktheatre-sw.co.uk/playbacktheatre/ (accessed 7 September 2022).

Music

The protests were also noteworthy for their creative nature, especially in urban centres like Beirut and Tripoli. An early protest in Tripoli went viral when a local DJ started playing music and the demonstration turned into a dance party while protesters across the country engaged in the traditional *dabke* dance at protests. Music was a constant element, ranging from Lebanese classic songs to new hits about the revolution to variations on 'Baby Shark',[3] while sound was also ubiquitous in humorous chants and cheers. Even in difficult moments, like the day when Hezbollah militants destroyed and burned the demonstrators' tents, people used song to come together. As one demonstrator remembered, 'When they came and destroyed the tents, people came together that same night and were singing songs. It was therapeutic and soothing, even my personal emotional wounds were closing, sitting there listening. It was so open and so peaceful, just listening to the music and sharing the positive vibes' (interview with author, 2020).

Storytelling

Storytelling and narrative have been central in FFP's work since before the revolution. Former combatants first go through a deep personal biography process to unpack and understand their own stories and experiences. They then share these stories through a variety of public fora, from documentary films to community dialogues, so that others might learn from their experiences. Storytelling and narrative were integral to the revolution as well. As Christina Foerch,[4] a biography trainer and board member with FFP explained, 'Humanity relies on storytelling. It's deeply rooted in human culture and is found different formats, even a TV series, or commercials, are forms of storytelling. Ultimately you can relate almost all of human culture to storytelling. And of course, stories and narratives give identity to a revolution' (interview with author, 2020).

During the revolution, FFP members engaged with other civil society actors, including NGOs, academics, artists and activists, to tell their stories in public dialogue. According to FFP members, they sought to use their stories of involvement in the civil war to influence today's activists to pursue social and

---

[3] Available online: https://www.youtube.com/watch?v=BNd2im6zYno (accessed 7 September 2022).
[4] Participants' names used with permission unless otherwise noted.

political change through non-violent means. Accordingly, their stories had two main target audiences. First, they sought to influence the younger generation of activists, many of whom were distrustful of civil war–era combatants whom they blamed for the current corruption. As Christina noted, 'Storytelling is very important in the intergenerational setting. Storytelling and story sharing are important tools for social cohesion, for social healing and change. It's important for youth to hear from ex-combatants, that at least some of them regret the past and support the current revolution' (interview with author, 2020).

Second, FFP sought to reach out to individuals who rejected non-violent means, resorting instead to violence or militancy. FFP members believe they are in a unique position to reach these audiences because they participated in violence in the past as well, thus giving them more leverage than a typical NGO worker or peace activist. As Christina continued:

> A group like FFP has an important role to play in shaping narratives and keeping a protest nonviolent because of their own lived experience of engaging in violence to reach a political or social change. Not only FFP members, but especially FFP members, because they have lived violence in person and experienced that it did not lead to the change they wanted. FFP can show empathy to these violent protesters and say we understand you, but we can tell you that violence is the wrong way, because violence leads only to more violence. So our ex-combatants are the best ones to tell the story because they've been there. (interview with author, 2020)

This decision to reach out to those with differing opinions or means represents a challenge for activists. Should activists focus on mobilizing those who agree with them, or reaching out to those who disagree? As Christina explained:

> There was a minority who couldn't identify with the revolution and tried to destroy it with violence. And we asked ourselves, how can we reach out to this other side? Because obviously there are many. But we never really found the answer to that. When you have your projects, do you try to reach out to those who are hard to reach, or those who are reachable? (interview with author, 2020)

Engaging with hard-to-reach groups and individuals, especially those involved with violence, comprises much of FFP's longer-term work, beyond the public spaces and protests of the revolution. These longer term processes of outreach, mediation and intervention also represent a sort of resistance but one that is embedded much more in everyday engagement and relationship building. Even in this work, however, storytelling and narrative are integral.

As one FFP member described, 'You start with individuals, then they find their ways to organizations, and then it becomes a bigger narrative' (interview with author, 2020).

Theatre

The concept of narrative and storytelling for mobilization has always been central for Laban in the theatre space as well. Farah Wardani, the founder of Laban, described how stories are crucial well before a revolution for building community solidarity and facilitating mobilization:

> We use the concept of the 'red thread,' the thread that we believe is present in every performance; it is a line that brings everything together. The heart of it is the community; people tend to be more empathetic with others than with themselves. People will often blame themselves, but when they hear someone else's story, they see it as a social problem that needs to be addressed, so it's not a matter of something between me and myself, it becomes a story of the community and it mobilises people to different causes. Some stories resonate more or less with different people, but they create empathy and the need to take action. People then move from being a witness to being more active and from the theatre to other places. (interview with author, 2020)

Indeed, Laban members saw the role of theatre as more crucial in creating spaces to prepare for and process the revolution rather than being central in the revolution space as itself. As Farah explained, 'In the past, we were trying to create an alternative space with our theatre. But now the citizens didn't need the alternative space, this was already happening in the city; public spaces were belonging to the people for the first time' (interview with author, 2020). Accordingly, Laban opted to support the revolution not through large theatre manifestations but through providing spaces for activists to reflect on the experience and process their emotions. In relation to the covert-overt spectrum, Laban's theatre-based approach helped activists process their overt expressions of resistance in more covert methods of reflection. As Farah described, rather than the usual model of starting with an individual story and linking it to broader social issue, they focused on doing the inverse:

> Usually we make the story a social event with public and political aspects; stories aren't just for the teller, but are communal. So if things are open now, maybe let's offer the space for the individual stories to come out. We were seeing that

some of the citizens were overwhelmed and feeling the ground was shaking beneath their feet, so we tried to offer a space to go from the big to the small, asking questions like, how are you going through this experience as a person, how is this affecting you as an individual? So we tried to be minimalistic in our movement and our choice of words, and focused on how to take a big story, and make it personal. (interview with author, 2020)

Similarly, Laban sought to provide spaces to capture stories of the revolution to ensure that the ideals of the moment were preserved. As Farah explained, 'We try to shape the collective narrative, the story of this generation and this revolution. A collective story must emerge, and we must play a role and keep the stories alive for future generations and show that we were not silent' (interview with author, 2020).

Throughout the 2019 revolution, FFP and Laban reflected on the relational nature of resistance, using stories and theatre to influence listeners and audience members to inspire greater social change while also using the emerging collective narrative to affirm and clarify individual experiences. Both also exhibited key elements of everyday resistance, including adapting a repertoire of techniques and approaches, and using methods that were temporally and spatially situated but not bounded. Further, their narrative and theatre approaches allowed for variance along the spectrums of individual-collective reflection, covert-overt expression and process-outcome orientation. Further, the sound and arts-based approaches facilitated a necessary blurring of lines between resistance, resilience and reconciliation, creating a thread between their past and future initiatives.

Beyond resistance: Resilience and reconciliation in Lebanon

The efforts of both FFP and Laban in the 2019 revolution were built on a strong foundation of community engagement that focused more on resilience and reconciliation. In this section, we focus specifically on how, in the years prior to the revolution, FFP and Laban used Playback theatre as a creative space for facilitating reconciliation between former combatants from various militias and sectarian groups from Lebanon's civil war, as well as victims, survivors and the general public. This creative space is especially crucial in Lebanon in the context of political elite 'amnesia', the related lack of educational curriculum about the war and general social-cultural taboos about discussing the past.

Indeed, many community members are hesitant to discuss the war for fear of reigniting tensions or reinforcing sectarian differences. However, we find that interactive storytelling via Playback theatre provides a notably constructive rather than destructive method for remembering. Rather than calling for revenge or punishment, this approach facilitates transitional justice by allowing for intrapersonal healing, interpersonal empathy, community-wide dialogue, and at times, reconciliation.

Context: Post–civil war Lebanon and forceful amnesia[5]

The Lebanese civil war, fought along sectarian lines, lasted for 15 years from 1975 to 1990, resulting in the deaths of over 120,000 people and the emigration of nearly 1 million people from the country. The Taif Agreement that effectively ended the conflict brought an end to the fighting but left communities divided (Haddad 2009) due to the lack of a meaningful framework to implement transitional justice components. As Craig Larkin (2010: 631) explains:

> War traces, in the forms of sites, absences, and narratives, have become normalized in everyday life, impacting spatial patterns, social encounters, and self/other perceptions. Amidst times of political instability and heightened tension, such spaces of imaginative connection and shared trauma are not only strengthened and reworked, sustaining prejudices and sectarian/political differentiation, but also offer protection, communal solidarity, and a sense of belonging.

Rather than moving on from a divisive narrative, Larkin explains that new Lebanese generations repeat the same fear and prejudices which make social fragmentation more likely as different communities hold the past against one another. As Eduardo Wassim Aboultaif and Paul Tabar (2019: 109) explain, 'The Lebanese case shows that communities often have two contrasting discourses of communal memory, creating in the process an internal and an external "other"'.

In this context of state-level reluctance to engage with transitional justice, civil society organizations have led efforts in dealing with the past. Many organizations, like Laban and FFP, use elements of art and oral history to document or discuss

---

[5] Andrew Mikhael (Queen's University Belfast) contributed to the research on Lebanese politics in this section.

the civil war. However, FFP and Laban's collaboration is unique in involving former fighters alongside victims and civilians and in using Playback theatre and narrative storytelling as a starting point for reconciliation. In the remainder of this section, we examine the impact of Playback theatre on the storytellers, performers and audience members. We find that the participatory and artistic nature of Playback theatre allows for the unique explorations of difficult stories that can create spaces for conversation, healing and, at times, reconciliation.

Intrapersonal: Reckoning with the past

Participating in Playback theatre can be personally meaningful for those who have experienced conflict, whether as combatants or as civilians, by providing a safe space to share stories and acknowledge the past. As Farah (the founder of Laban) explained, 'You feel safer because your story is being accepted by many people. The performance is very inviting, very acknowledging. Because your story is being acted, you are sure it will be listened to, it will be accepted' (interview with author, 2019).

Telling one's story and having it acknowledged can be a first step in personal healing and even transformation. As Gaby Jammal, an ex-combatant and member of FFP said, 'It is a form of healing, of transferring from a place where you belong to the past and saying, no, you have to take action. If you don't like your actions from the war, if you regret it, then it should be your mission in the next part of your life to address that' (interview with author, 2019). Likewise, Playback can provide a medium for dealing with unaddressed trauma from the war, specifically because it is rooted in emotion and expression rather than words. As Farah commented:

> Arts and theatre specifically are the key tools to address [trauma], because while talking is helpful, it will not lead you to anywhere. You need to embody the feeling. Language can't express everything. This explains why we use hand gestures all the time, facial expressions, because words alone can't tell what we want to say. This is why Playback is very helpful. It will help ex-fighters and people in the audience sitting and witnessing a story being played back to actually see the trauma moving, the story moving, the hate moving, all the feelings moving and actually coming to life in a heightened dramatic event. At the same time, the fact that it's a dramatic event is a kind of safety. (interview with author, 2019)

Playback can also help ex-combatants and others who have experienced war create a distance between themselves and their story, which Farah saw as an important part of therapy. While there is always risk that the process might 'open wounds' that may not close immediately during or after the performance, Laban actors are trained in dealing with sensitive topics and do their best to contain the performance to minimize further harm or pain.

Interpersonal: Encountering, humanizing, empathy, storytelling and dialogue

Intrapersonal healing is further underscored by the interpersonal nature of Playback theatre that allows for engagement between the story sharers, the actors and the audience. As Younes, a Playback actor, explained, 'In Playback, you experience the intimacy of the dialogue between the tellers, the performers and the audience. This intimacy helps channel the gravity of the war stories and create a lasting impact on the audience' (interview with author, 2019). Crucially, Playback provides a safe space where individuals from different lived backgrounds and sectarian backgrounds can come together. As Younes said, 'We have ex-fighters, civilians, refugees, victims … Civilians understanding ex-fighters. Ex-fighters understanding other ex-fighters. Civilians understanding civilians from the other side. It really bridges and tries to bring people together' (interview with author, 2019).

By bringing people from different backgrounds together, Playback can help in (re)humanizing and reducing 'othering'. As Samira, an audience member, commented, 'It's easy to view the war as something in which you were made a victim by people who were not people, but once you break down the fighters as human beings with dimension and complex thoughts, it's like, oh no, because you see the human perspective in the details of their stories' (interview with author, 2019). This humanizing potential was important for former combatants sharing their stories as well; as Gaby stated, 'I want people to know that ex-fighters, especially those who have made a transformation, although we participated in the war, in the end we are humans. So, they should look at us as a people who are trying to help prevent another tragedy of war' (interview with author, 2019).

In focusing on personal stories, Playback helps reduce feelings of 'us' versus 'them' or 'right' versus 'wrong' and encourages instead empathy and recognition of shared experiences. As Rana, another audience member said, 'It's our stories that make us human. Our stories are much more common than we think. The effect of stories overall is that they bring people together. People can sit down and realise, I've been part of something similar, or I've experienced something similar' (interview with author, 2019). Finding these commonalities can be an important step in transitional justice processes. As Younes commented, 'Bringing people closer around their own stories helps build bridges and assists in the reconciliation process …. There is a dialogue happening, and in the shared emotional experience that happens during the performance, there is some kind of collective memory that is being created, a kind of bridging' (interview with author, 2019).

Community and society: Constructing a bridging memory

While Lebanon has not completely ignored attempts to tackle the past, war exhibitions and memorization have been mostly the sole domain of civil society, who have used different techniques to promote a grappling with the past. From photo exhibits, traditional theatre and films, it was clear from our participants that in the highly politicized atmosphere in Lebanon, where state-sanctioned amnesia is present, arts were the primary means to engage with topics that are considered impermissible in public. As Farah explained when describing theatre events in Lebanon, tools like theatre and art have managed to 'break a lot of taboos that were untouchable'. However, as Farah continued, in the decades that have passed, 'The civil war topic is still a closed topic. It's a black box that no one can address and talk about properly in Lebanon' (interview with author, 2019).

Breaching the topic of the war publicly in Lebanon is a sensitive proposition due to the divided communal understanding of the past. Most often young people learn about the conflict in the privacy of their homes, in which the concept of martyrdom is especially rooted. As a result, intergenerational memory of the conflict occurs, with 'good' and 'bad' roles assigned, which makes breaking down the narratives even more important; Playback allows for the expression of those communal myths to be unpacked. As Gaby explained, the Playback process can bring to light the previously closed topic of the civil war: 'The discussion that happens in the session is important because [the youth] didn't dare before to talk about [the war], and because they heard only one side of the story through their families' (interview with author, 2019). The technical application of the Playback process helps create a sense of dissociation from social stigmas; as Farah details,

'The deconstruction of hero narratives, breaching taboo, interacting with the other ... the drama tools help this change' (interview with author, 2019). By using actors to retell stories, the memories are detached from the former fighters themselves; as Younes explained, 'They can "watch" their story from outside, to reflect, reminisce and reconnect with a fresh perspective ... the performance is one of a safe space for them to express themselves freely, with no judgment' (interview with author, 2019).

In confronting the memory of the war, the Playback practice is designed to open up discussions that do not occur in other civil spaces, and, as Samira explained, to offer an 'effective way to reach people and to leave a kind of print on any person who has never heard about the civil war or the post-civil war' (interview with author, 2019). Importantly, the stories expressed in the events provide valuable lessons on violence-to-peace pathways that can assist other groups who are currently in conflict. For example, Samira pointed out how Syrian attendees of the events were able to recognize similarities to their own conflict: 'Syrians who are dealing with it now, they wanted to know more, how the Lebanese people dealt with it and how they can avoid Syria becoming what is now Beirut' (interview with author, 2019). In this way, the Playback space allows for comparisons to be drawn across conflicts and to make audience members aware of some of the potential pitfalls that can occur in the post-conflict landscape.

The events can also create spaces of agency and meaning to help work on unresolved issues for the former combatants taking part, assisting them in gaining acceptance. For some, retelling their story helped de-stigmatize their place in Lebanon, particular because they now warn against the use of violence; as Gaby explained:

> I was trying to find a place for myself as an ex-fighter, meaning I concentrate on my personal experience to tell people 'if you are thinking to go to war, don't try to experience the thing I already experienced.' People listen to us; we notice that ex-fighters like us, when we go to the street and talk to other ex-fighters and other people, we are accepted, people accept us. (interview with author, 2019)

The addition of the former combatants helps add to the authenticity of the reconciliation aspect of the Playback events because they were personal witnesses in the front lines of the conflict. The presence of former combatants who have reconciled their past involvement in the war and committed to telling their story create a powerful narrative that lends credibility; as Farah explained, 'There is credibility that you're addressing the topic properly because you have ex-fighters on board, ex-fighters who went through a proper reconciliation process and who are willing to create change; people who went through war and who regret it now' (interview with author, 2019).

By adopting the medium of Playback, FFP and Laban have found a way to create a space that allows those who attend to exercise their own agency while learning about the past without, as Farah says, 'preaching …. We facilitate it but because it is improvised and people just come and tell their stories, people feel it and their emotions really come out. It's an effective tool and it's artistic' (interview with author, 2019). In addition, the inclusion of female former fighters in the performances helps construct a narrative in Lebanon that is most often male led. As Farah explained:

> When you actually hear stories of fighters who are women, and you can sense the feminine side of their personality, it's very emotional but you can feel the guilt and regret, not from social shame but out of a conscious decision. It's very enlightening and what young girls need to see, that women can make mistakes but come back to the image they want to restore about themselves. (interview with author, 2019)

By providing a safe space for intrapersonal, interpersonal and communal dialogue, storytelling in general and Playback in particular provides a restorative approach for confronting the past in a constructive way, especially in the absence of formal channels or mechanisms. As Younes said, 'Whenever we create an impact by opening a safe space to talk about those topics, it feels like we are taking a step forward in making Lebanon a better country for its people. This safe space is also a space for dialogue, where everything repressed can be discussed' (interview with author, 2019). Crucially, Playback provides a space to discuss the past in a sensitive, regenerative way. As Younes continued, 'It is a technique of remembering the war, keeping the war memory alive but in a way that will not make us fight again, that is the difference, trying to reshape the memory into something that is more constructive and less destructive, less inciting to conflict, trying to talk about our personal traumas or personal stories' (interview with author, 2019).

Indeed, by rooting performances in personal stories, Playback helps ex-combatants, victims and civilians move towards empathy, and sometimes even towards reconciliation. As Gaby stated, 'By telling our stories, we are urging people to express themselves in any way, without any prejudgement. Because the first step towards reconciliation is to be honest and start talking' (interview with author, 2019). Nassim Assaad, another ex-fighter, agreed, saying, 'The idea of reconciliation is to create new ideas and peaceful ideas, not the old ideas and stories of fighting and killing, but talking with an open heart' (interview with author 2019).

The spaces for transitional justice and reconciliation created by FFP and Laban are especially important in a context like Lebanon. The lack of engagement with the past affects not only those who lived through it but subsequent generations as well who have inherited the tensions but lack knowledge of the particular events. As Younes explained, the current governance helps reproduce transgenerational trauma across generations: 'The feelings are transferred, which creates a society full of tension but without any reason. That's what we're trying to deal with and to address' (interview with author, 2019).

FFP and Laban's merging of Playback with live testimonies is just one form of arts-based interventions to challenge the state-enforced amnesia over the Lebanese war. Both organizations fill an important gap in personal and community understandings of conflict; as Younes describes, 'All these erased memories were being shunned by the government and by the system we live in. It is very satisfactory to hear stories that are basically being systematically erased' (interview with author, 2019). FPP and Laban see themselves as fulfilling an important role in trying to push back against political and social inclinations to enforce a state of forgetting. However, more than simply a recovery from previous conflict, FFP and Laban are also using storytelling to help prevent further conflict. In this way, interventions like Playback contribute to a form of transitional justice that addresses not only the past but also the present and future.

Conclusion

As the Lebanon case study illustrates, resistance takes many forms, even within a single context. However, the sound and theatre-based approaches of Laban and FFP, both before and during the revolution, reflected the relational, everyday and spectral notions of resistance that we identified across the case studies. First, the relational nature of resistance was evident in the way FFP and Laban used stories and theatre to confront state power dynamics, initially by challenging the government-imposed amnesia on the civil war and then by using theatre, music and storytelling to confront the state during the 2019 revolution. In both cases, in accordance with the typology introduced earlier in the chapter, the actors and storytellers essentially functioned as activists, building solidarity with observers/audiences, while contributing to a broader movement challenging the target/state.

Second, FFP and Laban's activities both before and after the revolution reflected elements of everyday resistance. Though still coordinated and public, they employed a culturally relevant repertoire of storytelling-based tactics that were not overtly confrontational and created spaces for stories to transcend time and space through their resonance with the audience members.

Third, the sound-based practices used by FFP and Laban reflected the spectral notions of resistance. On the individual-collective spectrum, they managed to take individual stories and connect them to a wider community's experiences, weaving collective narratives out of individual reflections and allowing for both intrapersonal and interpersonal change. On the covert-overt spectrum, their performances ranged from small-group settings in a closed theatre space to public dialogues in the open square of the revolution. Regarding the process-outcome spectrum, the objective of social change was continuously coupled with the aim of personal transformation through the process of storytelling and theatre, and Playback in particular. Finally, and crucially, the activities of FFP and Laban illustrated how sound-based 'resistance' is inseparable from building individual and collective resilience and creating spaces for intrapersonal and community reconciliation.

The Lebanon case study is not unique in this regard. As the complement of our research in Brazil, Israel-Palestine and Northern Ireland illustrates, sound-based resistance is diverse in both its form and its impact. But sound, music and storytelling are a part of almost all conflicts and resistance movements. These techniques enable participants to quite literally get their voices heard through everyday practices in ways that challenge the status quo and cultivate new spaces for collective empowerment and community dialogue.

3

Resilience in creative practice in a post-conflict context: Musicians Without Borders

Fiona Magowan

Resilience is a key factor in recovering from the destructive effects of intercommunal conflict, and it is uniquely bolstered by community music-making initiatives, such as those employed by Musicians Without Borders (MWB), which seek to build peace in post-conflict contexts.[1] A 'resilient community' is one that can recover from the catastrophe of civil war, with a key emphasis on healing communal divisions and creating shared society, shared space and shared culture. MWB's work clearly holds these goals as paramount, and they utilize both verbal and nonverbal music-based approaches to develop interpersonal, empathic connections, through which meaningful reconciliation can be based. Another focus of MWB's work is recovery from trauma at the individual level, rebuilding confidence, developing leadership skills and regaining trust. Music-based approaches are also utilized to affect this transformation. In MWB's training programmes, these societal and individual foci are complementary, with overlapping music-based approaches. However, unresolved tensions between differing conceptions of resilience at the societal and individual levels reverberate in a closer analysis of MWB's activities.

In MWBs' training programmes, it is not clear whether individual transformation is considered temporally prior to societal change or concurrent with it, or indeed whether one conception of resilience might be considered primary over the other. Rather than view this dilemma negatively, while it contains the potential for contradiction, we argue that it should be understood

---

[1] Musicians Without Borders is an international organization, established in 1999, that is working across the globe to use music in peace building for healing, reconciliation and social change. MWB runs programmes in disparate parts of the world that have experienced or are still undergoing conflict. These include music, dance and theatre programmes for refugees or for those who have been displaced by forced migration. The areas of MWB's focus include the Middle East, Europe, Central Eastern Africa and Central America.

as a valuable form of *creative tension*. Thus, rather than erase one conception of resilience (individual or societal), or subsume one into the other, we show how these differentiated conceptions of resilience coexist in an antinomous, productive creativity through distinct yet overlapping sonic and music-based approaches and their affective resonances. In this music making context, affective resonance refers to 'the dynamic entanglement of moving and being moved in relation' (Mühlhoff 2015: 1001), which will be shown to increase resilience among participants.

The chapter begins by outlining the philosophy and approach of MWB, setting this in the context of theoretical analyses of some key concepts that underpin the creation of resilience in their work: trust, empathy, inclusivity, safety and affective resonance. We then explore the outworking of these dynamics across a range of music-making examples as discussed by trainees and facilitators to demonstrate the effectiveness of the techniques and the challenges that they present at both individual and societal levels.

MWB's Approach

The music-making strategy of MWB is primarily a means of generating group cohesion through socializing and empowering individual voices to be heard and to take responsibility for inclusive leadership via 'affective resonance', as much as through discursive critique. Their approach to training facilitation follows Christopher Small's concept of 'musicking' as everyday empathic empowerment. Small (1998: 2) notes that musicking is 'something people do', and in the participatory and improvisatory context of MWB' training, this means that the participants 'have an important and acknowledged creative role to play in the performance through the energy they feed (or fail to feed), selectively and with discrimination' back to the other participants (7).[2]

In our analysis of MWB's practices, we highlight the role of sound, both verbal and nonverbal, as well as body movements, gestures and facial expressions. The visual is a complementary dynamic to the sonic in participatory musicking,

---

[2] The participants are 'musicking' in so far as they 'take part, in any capacity, in a musical performance, whether by performing, by listening, by rehearsing or practicing, by providing material for performance (what is called composing) or by dancing, together with all the dimensions that support these activities' (Small 1998: 9).

since aurality and listening are key building blocks that are visually reinforced. This approach avoids prioritizing an ocular and logo-centric dynamic over that of the aural and affective. As Steven Connor (1997: 7) proposes in the *Modern Auditory I*, 'The self defined in terms of hearing rather than sight is a self imaged not as a point, but as a membrane; not as a picture, but as a channel through which voices, noises and musics travel.' From a related perspective, in his book, *Listening*, Jean-Luc Nancy asserts the primacy of listening as a core element of being attentive. Listening is about 'hearing-understanding (*entendre*)' and 'actively orienting oneself towards sound' (James 2012: 63). To be listening is 'to be on the edge of meaning, or in an edgy meaning of extremity ... as a resonant meaning, a meaning whose sense is supposed to be found in resonance, and only in resonance' (Nancy 2007: 7). Thus, listening (together with the responses it elicits) is a primary mode of engaging 'affective resonance'.

In the complementary relation between hearing, listening and responding, we analyse a range of 'affective resonances'[3] to show how they build resilient responses. It is important to understand how musical practices – theories, instrumentation and creative approaches – underpin affect. Music is not simply a metaphor for ways of being, it is a process and practice that has real effects upon participants. In participatory music making, the feedback loop between MWB trainees and facilitators is designed to enhance empathic empowerment, especially increasing tendencies towards trust, empathy and inclusivity, which trainees feed back into their own creative practice outside of the MWB programmes.

Critical to MWB's musical enskilling process is the capacity to listen, to imitate and to take the lead in music facilitation. To understand how resilience emerges through processes of affective resonance, we studied a range of MWB sessions to understand the technical, embodied and sonic tools that facilitators use to engage listening and teach personal capacity building and musical interdependence. We examine their potential to strengthen relationships and transform interpersonal awareness, enhance confidence and develop mutual support. As we shall see, these qualities, in turn, become affectively sustaining where, for example, expanding consciousness around the ability to trust has the capacity to build faith in another person's trustworthiness. We begin by exploring these qualities through a range of nonverbal music exercises which promotes

---

[3] See also Robin James (2012) on 'affective resonance'.

freedom of expression, distinct from the impact of words and sentiments in lyrics. Then, we examine how singing and song composition can offer a moral educational approach to increasing sociability and developing an individual's judgement about trust through musicking.

Foundations of trust in improvisatory practice

First, we need to understand not only why but also how the dynamics of trust are foundational to the improvisatory approach that MWB take to their general principles of participatory music making. Trust increases senses of security, releases the individual from fear and anxiety and facilitates the development of identity, as psychologist, Erik Erikson (1963: 118–20), has analysed through eight psychological crises. An absence of trust is not just a sense of mistrust that entails 'suspicion' of an individual or particular system but it is 'a profound anxiety or dread inhering in an inability to trust anyone or anything' (Giddens 1990: 99, 100). Furthermore, in attempting to explain the condition or quality of trust, Giddens does not equate trust with faith, confidence or knowledge (Middleton 2018: 76) but rather sees it as an interconnecting force that resides somewhere in the middle (Giddens 1990: 34). An individual may have confidence in the expectation of how another will respond to the situation, enhancing or undermining a sense of trust, depending on how that action is undertaken. Trust then is invariably based on not knowing whether the actor will respond appropriately or not. There is a judgement about the contingent nature of action and interaction that is closely related to anxiety and which, as we will see in the process of musical training, shapes the nature of musicians' responses to expectations around facilitation. The implications of uncertainty and indeterminacy in trust mean that the individual becomes vulnerable in revealing their degree of confidence in another's actions. Uncertainty, for example, requires that an individual takes risks in improvisation, thereby placing responsibility on themselves for their actions. At the same time, they may anticipate support from the group, expecting that their trust may be rewarded by appropriate or appreciative responses.

Therefore, it is not surprising that for those who have experienced conflict, trust is inevitably a difficult process, which, Middleton (2018: 78) argues, should also include the analysis of control and coercion. He asserts that we need to move beyond binaries of whether trust is a state (following Erikson and Giddens) or a decision to act (see Gibbons 2001) and instead seek to analyse 'practices' of trusting (Yamagishi 2001: 121, 124). As Yamagishi argues, 'Trusting is a form of social intelligence', and there is a 'mutually reinforcing connection between the ability to predict correctly other people's actions and practice in actually placing your trust in others' (Middleton 2018: 79). The equation is perhaps most predictable where there is consideration of specific tasks expected – the roles, the intentions and the capacities of the actors and recipients of the action. In this analysis of musicking, we will uncover varieties of 'acts, processes, relationships and tendencies of trusting' (80). In approaching the tangled field of trust in improvisation, we need to recognize that this is a 'fragmented, dispersive, multifaceted practice which requires a singular approach, an openness towards individual combinations and interactions of ear and eye catching musical as well as extra-musical elements' (Cobussen 2014: 21). In this analysis, we will see how the vagaries of improvisatory musicking and its associated affective resonances are critical to enhancing understanding and positive interactions.

Music Bridge programme

The data in this chapter was collected during collaborative research with MWB's Music Bridge programme, a partnership between MWB and Cultúrlann Uí Chanáin, which ran from 2015 to 2017 in Derry/Londonderry, Northern Ireland. The setting for the Music Bridge programme in Northern Ireland's second largest city of Derry/Londonderry has a complex and highly contested history. This deeply divided society has segregated roots that hark back to the siege of Derry (which is celebrated annually by the Apprentice Boys of Derry) embedded in the imposing early seventeenth-century walls located on the west side of the River Foyle with the unionist and Protestant Waterside to the east. Derry city is predominantly nationalist and Catholic. In recent times, fault lines of sectarianism have continued to run deep since the 1960s, when Derry was a hotspot of tension over gerrymandering and the focus of a civil rights movement. This led to the Battle of the Bogside with clashes between police and Catholic protesters, culminating in Bloody Sunday when 13 civilians were killed on 30 January 1972. In 2010, the Saville Inquiry declared innocent all those who were killed by soldiers. Just a month and a day after the Saville report was released, Derry was awarded the status of UK City of Culture 2013. Given this history of division and conflict, as well as the mixed identities of participants in the Music

Bridge programme, MWB's approach to training in this part of the world does not draw upon specific musical genres but rather teaches a broad approach to music making which participants can apply in their work.

In Northern Ireland, the Music Bridge course took the form of a series of four-day long intensive training programmes held tri-annually during January, April-May and November over a three-year programme. Trainees would sign up each year for these training sessions according to their completion. Our research was conducted via participant observation, semi-structured interviews and a focus group[4] exercise in April-May 2017. Magowan and Donaghey[5] participated in a range of workshop training across years 1–3. During breaks in the sessions, semi-structured interviews were held with all participants who were attending each of the year groups to explore their background in music facilitation, the motivations for their participation in the Music Bridge programme and how facilitation inspired and fed into their own music practice. We aimed to interview the cohort of 16 trainees, as well as five Music Bridge facilitators and an administrator. Due to some absences at the sessions, we interviewed 12 trainees, 3 facilitators and the administrator. The interviews revealed a wealth of motivations for choosing different facilitation exercises in their practices, as well as a diversity of approaches to shaping facilitation. In addition to understanding the rationale for facilitation, we sought to understand how the methods used enhanced modes of resilience and indeed resourcefulness as facilitators focused on understanding how MWB techniques developed the practical and somatic capacities of trainees.

Structuring improvisation: Expanding nonverbal safe spaces

Trainees were taught how to structure improvisation from the outset of the course. Often a morning session might begin with an introductory song in which an

---

[4] The semi-structured interviews were bolstered by focus group reflections on a recorded creative montage generated from materials gathered at the first training session in April, which was played back to the participants in May. Surveys were also conducted at the same time, and these responses are analysed in Magowan and Donaghey's Music Bridge and Training of Trainers 2017 Report, January 2018. Available online: https://www.qub.ac.uk/research-centres/SoundingConflict/FileStore/Filetoupload,885080,en.pdf (accessed 17 November 2021).

[5] Dr Jim Donaghey was a QUB Research Fellow on the Sounding Conflict: From Resistance to Reconciliation project from 2017–20. He is currently Research Fellow at Ulster University and PI in of an AHRC Early Career grant 'Failed States and Creative Resistances: The Everyday Life of Punks in Belfast, Banda Aceh, Mitrovica and Soweto'.

individual is invited to respond to the facilitator by singing 'hello' to a simple melody ending with their name and passing the activity onto the next person, which, as the song progresses, accompanies the singing with individually identifying body percussion or movement. By creating direct musical interaction and bodily imitation from the start in both verbal and nonverbal expression, the facilitator then moved on to working on nonverbal interactions through more structured mirrored body movements and dance actions performed in pairs. From the outset, the trainee was brought into a welcoming arena in which they could expand their imaginative horizons and develop collaborative interaction. In a specific MWB training programme, one facilitator explained to a community musician how he 'often started a session crawling on his hands and knees to defuse fear, provoke laughter and foster a sense of trust with new participants' (Aissa 2017: n.p).

Nonverbal communication was also key to the start of each morning's work at the Music Bridge programme in Derry/Londonderry. Seated in a circle, the facilitator would start by sending a body percussion rhythm or vocal sound around the group. This would gradually be elaborated upon with more complex rhythms, and then the sound would be passed across the group to another member tagged through a look or a gesture. These nonverbal games also operated by imitation and development, whereby, for example, a facilitator would start with an action or movement copied by others which would then be elaborated into a new exercise with accompanying vocals or sounds. As one facilitator noted, 'I open the musical vocabulary to all sorts of sounds, all sorts of noises where everything goes and we can be creative with sounds and it doesn't have to be taught.' She went on to explain how this idea needs to be embedded from the outset. This technique impacted upon the trainees in promoting safety and inclusivity. As a trainee noted, 'I'd be prancing about like a two-year-old, tearing up bits of paper and singing stupid songs and making stupid noises and laughing, an awful lot of laughing. So, breaking down those inhibitions, creating a safe area.'

Safety is not only embodied in the nature of the space that is created in which trainees learn to relax their inhibitions, test boundaries and generally be silly, but it is also important for facilitators to delineate the field of interaction in which trainees can make mistakes without entering into self-blame or criticism of each other. One trainee noted:

> I think a great part of that is the ability [for facilitators] to laugh at themselves as well, when they make mistakes. It's not that everything has to be perfect. Everything is acceptable really. When it comes to coming up with ideas, yeah,

we'll try that. See if it doesn't work, we'll just do something else. Giving people that freedom, it's very free. Very liberating.

For a facilitator, the question she/he must consider is when and how such 'mimetic excess' can happen and how to contain it to create sustainable processes of resilience. In Taussig's terms, mimesis happens not just in the recognition of cultural difference within but also in differences between modes of cultural awareness. He notes, 'Mimetic excess as a form of human capacity potentiated by post-coloniality provides a welcome opportunity to live subjunctively as neither subject nor object of history but as both, at one and the same time. Mimetic excess provides access to understanding the unbearable truths of make-believe as foundations of an all-too-seriously serious reality, manipulated but also manipulable' (Taussig 1993: 255).

Freedom in participatory music making is serious business. It is the capacity to engage in 'make-believe', to generate forms of playful excess beyond the realities of the everyday and its sonic politics. Yet, for trainees, becoming vulnerable means that there is always the threat of being found out or rather not being able to carry off the 'make-believe', if they are in fact too serious about being light-hearted. One Level Three trainee, Annette remarked:

> I just developed new skills, a new confidence, it challenged me in many ways …
> I was nervous about doing the [youth] workshops, very nervous. But, using the non-verbal communication skills was really, really powerful because you have young people, some of them don't want to be there … and one of the things I've done, the first thing that we learned was about being very serious about being silly. You know … you're very serious and you're conveying a message to the participants that you're doing something silly but you're very serious about this, so there's a purpose to this.

The willingness to let go and be propelled into sonic awareness of an intensely focused and highly sensory process is a move towards freedom. It promises an opportunity for transformation that would otherwise be undirected and impossible without the affirmation of the 'sonic marks [that] unconsciously guide our behaviour' (Augoyard and Torgue 2005: 3). Trainees, therefore, need to learn and understand how particular kinds of 'sonic marks' give licence for unfettered performance.

Improvisatory participatory music making, as conducted by MWB facilitators, involves mindfulness, empathy, mimesis and a performative habitus, which, in turn, provides a flow between sonic induction and assumption, sensorial awareness and interpretation. As we shall see in the following discussion about trainees' experiences, there are shifts back and forth between individual sonic, musical and bodily immersion and practical detachment that leads to a critical decentring and recentring of trainee's understandings about the impact of their nonverbal musicking skills at a group level. This shifting process between individual and group evaluation is perhaps most evident in how facilitators and trainees deal with exploring vulnerabilities in musicking.

Creative tension and productivity

Annette also spoke about how her training in the MWB workshops extended and increased her skills in working with children and teenagers in her external music making settings. Although at first she was concerned about the time commitment that the Derry/Londonderry MWB training expected, she quickly came to appreciate the value of the engagement. She noted:

> It was actually brilliant. It was really brilliant ... a friend had emailed me the link to the Music Bridge project ... [and] it has proven just to be the best thing, it just so complements the work that I do and it just made what I was doing so much better, it was way better. I learned so much from Musicians Without Borders.

However, some trainees also commented on the challenge of being put on the spot to 'perform'. The experiences of uncertainty and corresponding vulnerability were more difficult for some than others. By pushing beyond the bounds of their own comfort zones, the trainees also increased empathic affinity for one another. Stewart commented:

> I think you develop empathy for people just be being with them and laughing with them, being in the same boat. If you have to do something silly with your name and the next person has to do it, you know exactly how they're feeling. They're going, 'Shit! What do I do? What do I do?' It's the first time they had to put themselves in that position. So, you immediately go, 'Oh, he's panicking as well. They're just like me'. So, I think you automatically develop a sense of empathy when you're in there. And maybe feel a wee bit put in a spot, which we try not to do, but we still do it at these training sessions. You get put in a spot, you just have to do it. So, you can certainly empathise with everybody else.

In his book *Acoustic Territories*, Brandon LaBelle (2010: xiii) comments on this paradox in relation to how sound 'exists as a network that teaches us how to belong, to find place, to attune to others, as well as how not to belong, to drift, to figure acts of dislocation, and to dwell within experiences of rupture'. This process of engaging nonverbally requires developing individual and interpersonal coordination skills in rhythmic bodily movements, sounds and gestures. While these activities may come across as improvised play, they are well structured within a toolbox of musical activities that can extend and enhance the theme. As Annette noted:

> The whole aspect of designing a workshop, making sure that each exercise you do is linked to the one before and that you change your theme in a very structured [way, means] ... if something doesn't work, it's easier to improvise and try something different but if you're not organised and structured, and have your homework done, then things can very quickly fall apart, you know.

On the one hand, Annette commented that she found the MWB training programme intense over four full days' training that would leave her physically exhausted as the mental and emotional concentration required was 'constant and relentless'. On the other hand, she also noted how the act of stepping into the circle was 'a great release' and a 'great de-stressor' because there was no space for other cognitive processing to happen except the experience of being in the moment of creative response or proactivity, which Csikszentmihályi (1990) has referred to as an autotelic experience of 'flow'. Yet this flow of experience was not necessarily enjoyed by all trainees. Moyra elaborated on this dilemma:

> There was the thing where myself and T. had to be teachers and I didn't enjoy that at all, but it was really just because they were just proving a point. I didn't enjoy it, but it was beneficial because you realise that you know, this is something that you can come up against in your work.

Moyra found working collaboratively with others in the group easier than leading because she was more used to performing with prepared materials alongside others rather than improvising as a teacher. The challenge of being thrust into the leadership role raised insecurities, particularly relating to concerns around having appropriate qualifications. With a mixed group and varied backgrounds, many, though not all, trainees have had formal training in music therapy or degree qualifications in related areas such as drama or music performance. For those who preferred time to think through how they would lead a session, the challenge of being put forward into the leadership role was somewhat daunting. Nevertheless, the *creative tension* that arose for Moyra in the dissonance she perceived as to how she could move others through her musicking was offset by the benefits of having participated in a communal leadership capacity. In doing so, she enhanced her ability to shape the affective resonances of the group through the support of those around her.

Developing interpersonal resilience

MWB's approach to creativity is productive in its harnessing of fun activities that enable socializing, group cohesion and the ability for everyone to have their voices heard. In this situation, trainees variously nominate themselves to lead on activities or to be a part of the organism of creative practice. Nevertheless, as one facilitator, Emma,[6] recognizes, inclusivity and a desire to join in an activity do not in themselves generate empathy. In her external work with refugees, she notes that it can be an aggressive environment where the children are not always kind to one another. However, it is through a process of improvising, which develops reliance between one child and another, that it is possible to build levels of engagement that work towards transforming attitudes about aggression. This transformation operates internally and externally as children become more supportive to one another in their musical environment, and they take this with them into their external situations, enhancing their sense of community, thereby achieving greater cohesiveness than divisiveness. Emma noted, 'And it's transforming … the ripples of it are transforming the community quite strongly. Whereas it's been quite tough, you know, every man for himself, suddenly, you've got this group that is looking out for each other and helping each other and being supportive.'

From Emma's perspective, empathy that arises from this mutually transformative engagement is critical to any intuitive exchange in the music-making process. While it may be possible in one sense to make music without empathy, she notes that in MWB's work it is essential 'to be on the same … even more than the same wavelength'. That process is achieved through transformative musical imaginaries, whereby the musicking of another becomes the musicking of oneself, and the interconnectedness of the nonverbal and verbal reinforce this through otherness. Emma noted:

---

[6] Emma studied classical music, plays folk music and jazz and is employed in various session work. She has just moved into electronic manipulation and has been working with refugees in Edinburgh using the techniques of the MWB programme. She brings prior experience of similar techniques used in another capacity as a member of the Scottish Chamber Orchestra's Creative Learning Department.

> I can remember when I was studying classical music, I was still at school. And one of the tutors said, 'You need to try and put your thoughts in that person's head and think about it in terms of playing together – put your thoughts in that person's head and think how they think'. So, by doing those sorts of exercises and games, then genuine empathy is going to make genuine music, rather than four people playing at the same time, you've got a group that's playing one piece of music for one tune.

'Playing one piece of music for one tune' is an apt metaphor for the way that MWB work in terms of societal, emotional and transformational outcomes. As LaBelle (2010: xii) notes, 'Sound may create a relational space, a meeting point, diffuse and yet pointed; a private space that requires something between, an outside, a gap; a geography of intimacy that incorporates the dynamics of interference, noise, transgression.' As the refugee children's attitudes changed through sharing sounds of musical improvising, so did their potential for experiencing one another's 'geographies of intimacy', opening them up to new forms of interaction and new kinds of sonic relationality.

Such processes of learning to trust are also highlighted explicitly in terms of what MWB seek to achieve in their training. Yet there is a difficulty in pinpointing how transformations occur in trusting relationships. As Emma noted:

> If [transformation] happened on a profound level, then it happened. And they don't need to know what they want. Maybe, like you know, you hear sometimes people look back and go, 'Actually, there's that one teacher in school that believed in me, or that one thing that happened'. But they've not pinpointed that for decades that that was the turning point in their life. So, they don't need to know. They don't need to know that, isn't this amazing, we've gone through this incredible thing and I feel like a different person .... It doesn't matter. The fact that they have gone through that transformation and had that experience [matters].

The ability of musicking to rewrite memories with new interactions is key to interpersonal transformation. Participants reflect upon it, they will 'pine' for it, they will miss it and then look forward to the next event and remember just how much fun the previous one was.

Positive reinforcement in sonic effects

It is important to analyse how nonverbal interactions relate to emotional dynamics. As we have seen, sonic effects can generate excesses of emotion whose affective resonances persist in an individual's memory of the event and are prompted again by its repetition. Thus, there exists, 'between the sound and the sonic effect, not a relation of similarity but rather a set of mutual references between the sound, physically measurable although always abstract, and its interpretation' (Augoyard and Torgue 2005: 11). For MWB facilitators who are working with the feedback loop between the sounds employed and the generation of specific effects, the potential for multifarious and potentially conflicting interpretations is inevitable. In order to ameliorate any potentially negative associations to sounds or words, facilitators deliberately use sounds and lyrics that promote self-other affirming anamnesic values and avoid culturally divisive ones. Just as they use imitation between participants to create social bonds, so also repetition in body clapping, verbal sound games or instrumental rhythms in group improvisation can create 'an effect of reminiscence' or 'the often involuntary revival of memory caused by listening and the evocative power of sounds' (21). The efficacy of layering sound activities creates a density of the anamnesis effect 'merging sound, perception and memory. It plays with time, reconnecting past mental images to present consciousness, with no will other than the free activity of association' (21).

The relationship of sound to memory is also closely linked to the ability to recall a sound that is remembered but has not actually been heard. This modality of phonomnesis is most poignantly expressed in the ability to reproduce a melody through a process of 'inner listening' (Augoyard and Torgue 2005: 85). Furthermore, the value of sound in these musicking processes is in the ability to generate excess and at times euphoric or phonotonic effects that lead to the desire to repeat the action or activity in order to relive the experience.

Trainees also employ these techniques outside of the Music Bridge workshop settings in their independent music practices. Stewart, for example, explained how in his work playing at a prison in Scotland he was not sure what the impact of the music was at the time, though it was to become apparent later. His group played for an hour in the high security environment as inmates watched and clapped. It was only when one of the inmates emailed him some years later that he became aware of the impact that the band had had. The inmate wrote to him, 'Oh I was at your thing … And I loved it and I've taken up pipes … and started doing woodwork.' Much of the motivation for trainees to continue to make music in their amateur/professional settings comes from inspiring others to do new things in their creative capacities because they understand that the phonotonic effects they create are recalled by their participants through an inner awareness of phonomnesis much later in their lives.

MWB trainees are also taught to be aware of how interpersonal transformation occurs in their participants. This can be as subtle as seeing 'a sparkle in their eyes' as one trainee noted. Recognizing these subtle changes is part of the ethos of learning how to lead viscerally in the MWB domain, an experience that flows between facilitators and trainees. One trainee jazz performer reflected about the facilitators, 'They just have a lovely manner about them. And it's not that they tell you, you should do this, they embody this real, like, love and warmth. It's hard to put it into words, you kind of have to experience it.' She noted how their approach contrasted with hierarchical forms of musical learning and engagement. Instead, there is a generalized reciprocity of emotional affirmation and reinforcement of the best aspects of being human together. This affect inherent in musical bonding also reorients a cognitive approach to inclusivity to an embodied togetherness that encompasses not only empathy but also the broadest range of emotional dynamics of trust, respect, friendship, kindness, positivity and warmth.

Consequently, while critical discussions around the impact of trainees' musicking techniques are reflected upon in one-to-one analyses with facilitators who assess trainees' understanding of their skills, this reflexive practice is not part of the musical activities in the training programme itself. Facilitators employ this technique privately in order to ensure that it does not disrupt transformative outcomes from the musicking experience, since it does not in and of itself enhance practical transformation.

Inclusivity

In addition to sound and rhythm exercises and movement activities, MWB trainees also learn the skills of song writing and address issues of lyrical and semantic complexity. For example, in keeping with the improvisatory approach of MWB, Annette employs a range of song-writing techniques that engage children by using 'catchy, easy songs that are easy for children to sing within a range', teaching lyrics rhythmically and avoiding sheet music. Making music learning aural rather than textual enables children to be freed from the pull of the written words on paper to focus more closely on *how* they are singing with others, rather than on what they are singing on the sheet in front of them. She also brings fun and competitive dynamics into the learning process, inviting boys and girls to sing alternate lines for each other. She said:

> [The children] would make it their own and they'd have ownership of it. And they would feel that this was their own ... I found that by the time I got to the second workshop, I had realised, oh my goodness, these children ... this child ... was very shy, but this child realised that she had a skill and she could use it here. Wow, you know, and there were so many occasions when that happened ... I wouldn't allow the teachers to give out lyrics, song sheets. I said no. I just taught by repetition and you know – so what does that message mean – and just broke it down.

She also progressively taught youth about the importance of nonverbal interaction, explaining how it is used by MWB and why it is powerful in other countries like Rwanda or Bosnia. Thus, the children came to understand how, as Labelle (2010: xii) notes, 'An entire history and culture can be found within a single sound; from its source to its destination, sound is generative of a diverse range of experiences, as well as remaining specifically tied to a given context, as a deeper expressive and prolonged figure of culture.'

In addition to working with youth, the variety of exercises that trainees were taught in the Music Bridge programme also set them up to lead workshops in external adult programmes. In this context, Annette had the opportunity to work with sixteen ladies in Derry/Londonderry who had been together for about ten years comprising various age ranges, generations and personalities, representing a cross-section of various backgrounds from across the community. The group included some with restricted mobility and some with hearing impairments. They met twice for two hours a week, and Annette delivered a series of five workshops. Preparing for such a diverse cross-sector meant that she had no idea what would or would not work in advance. Rather, she drew upon the various exercises employed by MWB, tweaking and rewriting them as necessary. She commented, 'It challenged them, they found some of the exercises challenging but when I asked them, "would you recommend this workshop to your friend?" they said, "oh yes", or "are you glad you participated in this?" and they said "yes". Annette adapted the exercises to ensure that they would be inclusive of everyone regardless of ability at the same time as being mindful about how to elaborate on the training exercises. As she noted, exercises can be built up to allow the trainee 'to step out of your comfort zone and improvise and, you know, expand it and if you see something is working, go with it and develop it more'. Annette recognized that she needed to work on this aspect of releasing creativity and improvisation in the group in order to enable them to take ownership of an exercise, which, in turn, she noted, 'opens up their creativity ... and you can see them coming alive'. The transference of power in the process of creative practice facilitates the shift in the experience of ownership from one voice to another and thus offers a means of providing an inclusive egalitarianism.

Transformational resilience

Just as Annette noted how she had to adapt to cope with these changes, equally, other trainees identified changes in themselves and their fellow participants during the workshops. Stewart commented:

> I see small changes probably [in] everybody in the workshop. You see the changes in the leaders from training to training, from workshop to workshop ... doing stuff that you thought ... you couldn't have done that last time and I couldn't have done this before. Changes in yourself as well.

He further reflected on how there were those at music-making workshops that the Music Bridge trainees facilitated for a school in Derry who embraced the work at the start while others would hold back, reticent about participating. He noted how they changed during the workshops as he saw them develop. He commented, they 'put themselves forward and you see a certain confidence growing in them'. He explained how the trainees on the Music Bridge project also work with an autistic organization with eight children, and while trainees may not recognize the transformation taking place during their workshops, the teachers noted how their confidence had been enhanced over time. Stewart puts this improvement down to the fact that 'they're included and equal and taking part and enjoying themselves'. He reflected that though it may be a small difference, every difference is important in these children's lives.

These transformations are made possible by skilled facilitation in which facilitators recognize an equilibrium between '(1) being prepared and able to lead and (2) being prepared and able to hold back, thus enabling the group of individuals to discover the journey of musical invention for themselves' (Higgins 2012: 148). In these subtle shifts, sonic effects are a mediatory force for empathic transformation. This is particularly true of the mimetic effects of improvisatory activities which enfold others in the same space of action. Taussig refers to mimesis as 'the nature that culture uses to create a second nature, the faculty to copy, imitate, make models, explore different yield into and become Other' (Taussig 1993: xiii; see also 251, *passim*). The effect of this yielding into and becoming Other is the revelation of 'the sensuous moment of knowing [which] includes a yielding and mirroring of the knower and in the unknown, of thought in its objects' (45). As Josephides (2010: 168) also elaborates, 'This moment of knowing is not passive receptivity, but an active imperilling of the self that removes the mask.' Consequently, improvisation is a risky business precisely because there is the euphoria of coming to know the Other, but as we have seen that requires becoming vulnerable, laying bare *the potential to see oneself* in the critical gaze of the Other. In her work on Tamil refugees in Norway, Grønseth argues that 'self reflexivity only produces understanding when, by an experience-near and empathic approach one recognizes that "they" are as of "us" as "we" are of "them" and as such allows us to project our imagination into the world' (169). Josephides (169) suggests that while Grønseth did not intend to 'become the other' with those with whom she was working, she did want to 'share experiences on the borderlands through empathy and engagement'. Such borderlands are places of encounter, and, in the case of MWB, they are also musical borders of bodies and being that respond to the expressed desire to negotiate new forms of exchange and ways of interacting that are exaggerated through improvisatory practices.

As I have argued above, a safe space for musicking can be a potentially conflictual domain since it entails a relatively private space for the exchange of communication, but it is also a semi-public domain in which self-awareness and reflexivity are intensified, becoming the object of self-other scrutiny. Despite nonjudgmental rhetoric, the ability to negotiate musical freedom with appropriate interaction is a playful skill. Trainees come to recognize that safety does not necessarily mean remaining within one's comfort zone. In fact, the opposite is true, and the safe space of musicking is perhaps a misnomer since it provides an arena in which there is 'positive creativity, exploration, and future happenings, generating a safety without safety space in which to nurture participants' potential' (Higgins 2012: 153). Improvisation is a form of moral value in which participants are active agents in choosing their degrees and types of engagement. Discussing Joel Robbins's view of moral action, Jarrett Zigon (2009: 253) analyses how the moral lives of individuals entail both 'an unquestioned and nonconscious aspect, as well as a conscious questioning that allows for freedom and choice'. It is in the process of 'choosing between values that [participants] come to see their decision-making process as one engaged with moral issues' (Robbins 2007: 300). The fact that MWB participants most commonly allow themselves to become vulnerable and risk judgment means that they are choosing to adhere to a moral code in that performance context, which is less likely to be applicable outside that setting. The so-called safe space then can generate potential internal conflict for participants due to the fact that several 'moral-value spheres seemingly determine moral action' (Zigon 2009: 254).

If we apply Zigon's theory of multiple conflicting moral principles to musicking, taking the decision to risk vulnerability has three intertwined dimensions: (i) 'the institutional; (ii) public discourse; and (iii) embodied dispositions' (2009: 258). Normative expectations around musical learning in this participatory arena provide a template for moral action that drives the participatory response. These actions then feed into a broader dialogue that bond trainees as they feel their way through performative experiences. These unofficial discourses may be distinct from or contrast with the participatory moral code, but they also inform how moral expressions of participatory action are understood.

In addition, when we consider resilience in musicking, there are simultaneous dynamics of unconscious and conscious engagement, as well as self-management. Resilience is bound up in an ethical reflexivity that demands that each participant works on themselves to 'become reflective and reflexive about [one's] moral world and moral personhood and what [one] must do, say or think in order to appropriately return to [one's] nonconscious moral mode of being' (Zigon 2009: 261). As Zigon notes, 'Ethics, then, is a conscious acting on oneself either in isolation or with others, so as to make oneself into a more morally appropriate and acceptable social person not only in the eyes of others but also for oneself' (261).

However, it is when one steps back to consider what is happening in each of the three moral dimensions that potential disconnects or 'moral breakdowns' (Zigon 2009: 262) can occur. Resilience is expressed in the capacity to expand one's moral purview and its application by working through these problematic dynamics that trouble selfhood and interpersonal relations and which, in turn, enable 'the cultivation of embodied and nonconscious morality' (263). Participatory music making offers a key arena for the development of resilience through the work of moral personhood because 'ethics must be regularly performed' (263). In reality, as we have seen in the trainee exercises, this process engages with the revelation of vulnerable difference, which, if taken to an extreme, can constitute a moment of moral breakdown, but the moment of *creative tension* also offers the grounds for repair and resilient growth as the participant seeks to work back to performing with the group through a 'nonconscious morality' (263).

By highlighting processes of moral value and moral breakdown in musicking, we can understand how ethical performance shapes processes of both individual (neoliberal) and societal (ecological) resilience and how the musical context provides a unique setting of expectation, listening and creativity. It also offers a way of exploring being-with-others in nonverbal, emotional, embodied and empathic processes that are fundamental to the ethics of interpersonal and social transformation.

Inspiring ethical awareness through the 'acoustic politics of space'

As this chapter has elicited thus far, MWB training strategies effect individual and interpersonal transformations that can indirectly address broader societal issues. Although MWB's ethos does not engage head on with the traumatic effects of conflicts or their politics, their techniques are highly effective as they develop resilient responses among those who have been victims of violence. As one trainer noted, the ethos to 'embed nonviolence, connection, empathy, respect, safety and inclusion into the work transcends differences in political views, race, gender and so on'. Trainees take the moral and ethical values of MWB practice into their own compositional processes and music workshops beyond the MWB programme. Working outside MWB training programmes, they often need to consider how to address issues of resilience in contexts of protracted conflict, as well as post-conflict environments. Thus, they develop a sensitivity to and moral code of musical engagement that embodies and critiques what LaBelle (2010: xiv) calls an 'acoustic politics of space'.

To examine this process, let us return to Annette's musical journey which preceded her enrolment in the Music Bridge programme, and which, in turn, continued to inspire how she creates work in other contexts. Annette, like the other trainees, brought her own compositional and creative talents to bear throughout the programme, developing further critical reflections from MWB trainers and acquiring exercises which she could relate to her own practice. For example, Annette had composed a school musical production with script and resource booklets in 2011 which addressed climate change and urged looking after the environment, encouraging youth to deal with littering. The production for ages eight and upwards was accompanied by her inaugural CD of songs which later became an educational resource pack that she launched in 2014.

Annette's desire to educate people about littering was reflected in her passion for composition, and her work demonstrates the ability to play seriously, which brought attention to environmental details that impacted on school pupils. Her ability to amplify the inanimate resonates with 'the greatest artists of all time [who] had a knack for playfulness, for seeing the life inside of things' (Lederach 2005: 160). She was asked to write her first song in 1998. The songs were inspired by a network of connections. Annette's close friend asked her:

> Could you not write a song that will be so catchy that, you know, men would be whistling it on their way to work? She replied, 'I can't promise you anything'. Then continued, 'But I actually wrote the first verse in the shower. So, I dive out of the shower grab a piece of paper and write it down. It would be a year later before I completed it.'

She further explained how a late county councillor had organized a song-writing award for the Tidy Towns competition and was aware of Annette's compositional abilities. She encouraged her to write more songs, and the CD for the production was realized. While Annette had been inspired to create the first two songs herself, the next four were then commissioned by the Education Training Board for Dun Laoghaire County Council. In rolling out the production, Annette has worked successively with five councils through the Local Agenda 21 Funding initiative. Invitations also came from other councils, such as South Dublin County Council, who invited her to work with both secondary and primary school children on the launch of their Eco Week 2017 programme. This allyship meant that the councils became the centre point for distribution of the first packs. They also purchased subsequent packs, as well as contributing funds to two schools to put on productions for other schools. She explained how the MWB Music Bridge programme had given her time away from the stresses of the launch of her educational resource packs with all the demands of sales and invoicing, providing a sense of escape into a mindful experience.

The eight songs in the musical track environmental transformations and moral societal messages. The first one tells of the beauty of Ireland in that it is 'The Place to Be'. It is followed by the song, 'Don't Drop Your Litter', teaching respect for the environment. The story is set between two fictional sites, Millview Park and

Greenway, but as the musical unfolds, it becomes evident that the Millview Park residents are not as careful of their environment as those in Greenway. Nevertheless, the children are friends and spend time hanging out together. Millview Park becomes messier with broken glass, fly tipping and other waste evident in the surroundings, as evoked in the songs 'Dog Poo' and the need to recycle in 'the Bottle Bank Song' and 'Sorry', the latter apologizing for blindness to the state of the environment.

With the arrival of two tourists who are looking for directions, the youth point to the sign which has been graffitied to read Messview Park, and they suddenly realize that the signpost has been altered. As the tourists leave, the young people realize that Greenway is much cleaner and tidier due to the Tidy Towns volunteers being active, as evoked in 'the Volunteer Song'. So they enlist the help of a friend's uncle from Greenway to mobilize the Tidy Towns volunteers to help clean up Millview Park and in the process teach the young people about recycling and reusing items that are made of waste. They come together to host a fashion show with clothing made from rubbish to illustrate their changed understanding as they realize in the final song 'It's in Our Hands'. The songs embody processes of social exchange that come to be mutually transformative not only of the environment but also of the children's relationships with one another. The musical thus teaches children not just about respect for their surroundings but also about respect and humility in helping their neighbours.

The growing ethical awareness of environmental issues and the need to resolve the conflicts arising in the songs illustrate how 'sound opens up a field of resilient interaction which becomes a channel, a fluid, a flux of voice and urgency, of play and drama, mutuality and sharing, to ultimately carve out a micro-geography of the moment, the instant of hearing, while always already disappearing as a distributive and sensitive propagation' (LaBelle 2010: xii). The music facilitated a heightened sense of attunement to the environment and listening to environmental issues which changed youth behaviour towards those spaces. Therefore, it is not politics that has changed the town but attention to the acoustic politics of space that has enabled youth to become healers of what was once destroyed. As Lederach (2005: 162) notes:

> The greatest movements forward, when you look really closely, often germinated from something that collapsed, fell to the ground, and then sprouted something that moved beyond what was then known. Those seeds, like the artistic process itself, touched the moral imagination. To believe in healing is to believe in the creative act.

Intercultural empathy

Although the example above is about teaching children moral values, its political sentiments resonate with conflicts in everyday environments. In a related but different vein, one Music Bridge trainee alluded to how participatory musicking can engage other aspects of intercultural conflicts, when he reflected on how he found it was unbelievable that he never met a Protestant until he went to university. Addressing conflict issues around political divisions is not directly part of MWB's work, which rather aims to promote sensitivity in responsible leadership and creative interdependence. Stewart noted:

> If you get people laughing and having fun together, it's an enormous step between building trust and building understanding and empathy. It's kind of what MWB is trying to do, give people a voice as well. You've got a lot of these places where people don't have voices and they don't have any way of expressing their ideas or their individuality or their personalities. So, it will help people come out of that.

As we have noted previously, MWB work around the world in many conflict zones, and facilitators are careful to avoid specificities of language or to engage with political issues pertaining to conflict. In doing so, they carefully combine understanding and awareness of the political, yet without addressing conflicts in their work. One trainer elaborated on why this avoidance was necessary. He noted:

> "[If] we are talking conflict resolution, the implication is that I want you to stop conflict, I want you to feel differently at the end of it. And I think that already sets up the barrier. But if you say, 'We're going to make music together and actually you're welcome as you are and who you are ...' [then] what you think politically is not relevant in this space, although it will come up because it's part of who you are".

Thus, MWB's approach to peace is one which seeks to approach the psychology of trauma tacitly by providing new tools of embodied practice. As director of MWB Laura Hassler (cited in Burnard et al. 2018: 235) notes, 'To support processes of re-connection without identifying people by ethnic or cultural labels, Musicians Without Borders works to create a neutral musical space in which participants can both identify (themselves) and relate (to each other) primarily as musicians.' While MWB trainees take an approach to peace building that opens ways of listening deeply to each other's rhythms of music making, they do so by ensuring that their approach does not 'trigger' emotional trauma related to ethnic violence or cultural division. In achieving culturally sensitive, productive creativity, MWB have developed a highly attuned and precise way of negotiating and managing the complexity of 'acoustic territoriality' (LaBelle 2010: ix) without asking participants to verbalize how they deal with resonant emotions that underpin relations to place and which are mediated by listening to their environments.

This approach was evident, for example, in our brief visit to MWB in Palestine.[7] Even though the occupation of Palestine is ongoing, MWB still maintains an approach of facilitating personal and collective resilience through music, rather than advocating for a particular political viewpoint (resistance) or pushing participants towards conflict resolution (reconciliation). It should be noted that the current focus of MWB's work within Palestinian communities is in the Bethlehem and Ramallah areas of the West Bank, (though they have worked all over the region), but they do not engage in cross-community programmes between Palestinians and Israelis. Rather, the focus is on engaging with marginalized communities within Palestine, including refugee youth, women's groups (including victims of domestic violence) and children living in low-income or underserved areas. Programming ranged from one-off workshops to more structured programmes, usually consisting of similar methods mentioned above, including movement, expression, rhythm, body percussion and using paper and other simple materials to make sound and music.

Although the Palestinian context is significantly different from that of Northern Ireland, the effects on resilience discussed in this chapter were similar there as well. First, MWB provided a physical safe space for participants and one that allowed them to channel their emotions into something expressive. As one trainer noted:

> Music can make you feel things. Either you can express your feelings through music. Or you don't need to express your feelings, but you can go through your feelings. If you play happy music you can get happy from it. If you play sad music you can allow yourself to feel like maybe you're sad but you can have these feelings at that moment, when in daily life maybe you can't show that you're sad, or you have to be strong.

---

[7] The methodology for the comparative case study of MWB in Palestine took place in March 2017 and consisted of six semi-structured interviews with staff and trainers, informal conversations with participants and two weeks of participant observation of MWB programmes.

Another trainer commented, 'Music is multi-dimensional. It adds the emotional part. It gives you another opportunity to express yourself, to express feelings. Whatever feeling you have, you can express it without throwing a stone, without violations. You still can express anger, anything, happiness, sorrow, you can express it with music.'

Second, like the Northern Ireland programme, MWB in Palestine fostered interpersonal trust, inclusion and empathy through the process of music making. As one trainer said, 'When you play music together with others, there is automatically a connection. When we do it together, there's a connection, even if we cannot explain it or name it, there is something, that we were doing the same thing, so there is something between us. We need each other.'

Third, the MWB music programmes in Palestine facilitated intrapersonal resilience and healing. As one trainer commented:

> When I meet a kid who is newly in music, I can see that his attitude is not stable, I don't see stability in their lives, in their way of talking, their attitude, their way of thinking and communicating with others. After two or three years though, after engaging in musical programmes, I see a big difference in their personalities, in their attitudes. In their visions, how they talk, how they see life. How free their minds are, whereas before it wasn't like that. So music makes a big change for the kids I meet and work with.

Other trainers and parents made similar observations about children finding joy and a sense of purpose and dignity through the music programmes.

Conclusion

This ethnographic exploration of how MWB facilitators and trainees navigate the sonic and rhythmic effects of participatory musicking and address the complexities of diverse 'acoustic territories' offers some illuminating insights into processes of resilient action. As this chapter has revealed, processes of trusting involved in musicking depend to a certain degree on how musical practices are encouraged. In particular, we have seen that the analysis of trust needs to take into account the 'ways in which local systems of "social aesthetics" are diversely constituted and involve the deployment of "different orientating values" for the evaluation of "different kinds of performances"' (Brenneis 1987: 238). Critical reflection on these values reveals how individuals become motivated by musical listening and music participation, which constitutes a 'necessary condition for "ethical action"' (Hirschkind 2006: 13). In other words, how participants respond to the conditions of musicking depends upon their dispositions towards the mode of creative practice as well as the rules and expectations that the performance context entails. As conditions for trusting relations come to be fulfilled, trainees begin to develop a sense of community through what Higgins (2012: 141) refers to as 'the hospitable welcome that encourages participants toward creative music making and can produce lasting impressions on both community musicians and participants'. The types of exercises employed by MWB facilitators, together with their inviting means of communication, inspires increased social bonding as well as senses of '*coresponsibility*' (141), as trainees learn to develop creative freedom in a safe space that allows personal vulnerabilities to be expressed. As the boundaries of creative expectation expand to encourage new kinds of play, so also some trainees may experience the opportunity for risk taking as challenging to their senses of the musical self. However, as each moment of risk taking is rewarded by approval and reciprocation from facilitators and the group, so also trainees learn that the expansion of interpersonal energies can positively reinforce the limitless potentialities of play. Providing new spaces for such physical, mental and emotional transformation to occur thus entails deterritorializing established forms of creative practice, enabling alternative modalities of creativity, care, solidarity and ultimately resilience to emerge. It is thus within the indeterminate and liminal processes of rhythmic improvisation, dramatic play, song writing and free singing practices that the dynamics of affective resonance are felt, reinforcing the power of relationality, affording opportunities for interpersonal healing and social development.

4

Remediating relationships: Collaborative storytelling and conflict

Ariana Phillips-Hutton

'The world is full of stories because it is full of lives.' So claims Adriana Cavarero in the closing pages of her book *Relating Narratives* (2000: 143). This is also a sentiment that also echoes in the work of the composer/documentarian Mary Kouyoumdjian (b. 1983). An Armenian American musician born in California to a family that has suffered multiple generations of exile, many of Kouyoumdjian's compositions engage directly with the experiences of those caught in the middle of violent conflict. The origins of this stem from a desire to understand more of her own family history. Her parents sought refuge in the United States during the Lebanese Civil War (1975–90), a displacement and a conflict reflected in *2 Suitcases* (2010/2017) and *Bombs of Beirut* (2014), while a series of compositions including *Aghavni* (2008), *This Should Feel Like Home* (2013), *Silent Cranes* (2015), and her recent collaboration with Canadian Armenian filmmaker Atom Egoyan *They Will Take My Island* (2020) articulate the impact of the Armenian Genocide (1915–23) – from which her grandparents fled – on the Armenian diaspora and beyond. Still others of her works expose the results of contemporary human conflict: from the series *Children of Conflict* (2013–), which reflects on photographs taken by Chris Hondros of children in wars, to *I Can Barely Look* (2016), whose libretto is taken from children's responses to media-circulated photos of the Syrian refugee crisis. Kouyoumdjian's body of work is thus permeated by the stories of refugees and exiles. However, she does not compose merely to broadcast knowledge about the impact of violent trauma on individuals. Instead, Kouyoumdjian's music is concerned with a reciprocal musical storytelling and with the human relationships that are generated or altered through the telling of those stories.

As the other chapters in this volume demonstrate, Kouyoumdjian is far from the only contemporary musician captivated by music's capacity to unfold trauma, but her unflinching dedication to narrating experiences of violence and displacement and to granting marginalized people musical spaces in which to tell their stories make her work a particularly rich field for exploring music's remediation of human relationships. In particular, Kouyoumdjian's commitment to storytelling in and through music demands a radical approach to composition that blends field recordings and interviews with the production of notated music and the use of audiovisual media. The resulting compositional interweaving of fragments of speech with acoustic, electronic and prerecorded sounds is what the composer terms a 'documentary style' that deploys the sonic, technological and performative resources of contemporary music in order to faithfully communicate the real words and experiences of the real people involved.[1] This is in sharp contrast to the traditional narrative in Western art music of the lone creative composer sitting at a desk in his or her (but invariably his) garret. Yet it also suggests a model of creative practice that is different from the creative collaboration characteristic of group improvisation or other participatory practices common in musical activities dedicated to social transformation. Instead, it finds unexpected resonances with the principles of ethnography, particularly with the ethical principles that encourage sensitive and collaborative methods of research. Examining what I call here Kouyoumdjian's 'ethnographic' compositions reveals new aspects of the capacity of music to reconfigure human relationships. This, in turn, offers a new perspective on how the sonic art of storytelling is combined with musical performance to bring about social change.

I first came across Kouyoumdjian's music in 2016, when I heard the composer's *Bombs of Beirut* performed by the Kronos Quartet in London. Initially, my curiosity was piqued by the fact that I knew very little about the Lebanese Civil War or the city of Beirut, but as the piece got underway, the sheer force of the sound (a combination of live string quartet, processed sounds and archival fragments, including a missile bombardment in 1970s Beirut) in combination with the striking narrative grabbed my attention. A few years later, I got in touch with the composer to discuss her work. This led to a series of interviews and email exchanges with Kouyoumdjian and with several of the people with whom she worked to produce her pieces, *Paper Pianos: 'You Are not a Kid'* (2016) and *They Would Only Walk* (2020), including Milad Yousufi, Yamilette García and Buffalo String Works (hereafter BSW) Executive Director

---

[1] For more on the use of archival testimony in music, see Phillips-Hutton (2018, 2021).

Yuki Numata Resnick. In addition to interviews, I have also had access to several other resources, including videos of performances, a documentary film, critical reception of the works and Kouyoumdjian's own doctoral thesis (Kouyoumdjian 2021). This chapter draws on insights gleaned from those interviews about compositional and performance practices alongside analysis of these two works in order to argue that this 'documentary' confluence of oral history, ethnography, composition and musical performance provides a new model for understanding musically mediated social relationships. In addition, these pieces provoke a reconsideration of assumptions about music's connections to voice, identity and storytelling that shape the field of the sonic arts and social transformation even as they contribute to narratives of resilience among both those individuals and communities whose stories animate them and among their audiences.

This chapter therefore tells not one story but many. Stories that weave into and around each other and that speak to and with one another, even as their shifting relationships generate yet another, creating a still richer and more polyphonic story. What these stories tell is at once particular – how long does it take to fly from Burma to Buffalo? – and general – how do one's childhood experiences reverberate into adulthood? Some of these are stories we were told by interviewees, some we have merely heard at second- or thirdhand, and some we have created in order to tell them to you. In this telling and retelling, the stories illuminate the fullness of the world and its many lives, but they also make clear that none of these lives are lived in isolation. The stories we tell with the lives we live are stories of relationship sounded out with others. So I invite you to listen now and let the stories sing.

Playing Paper Pianos

In February of 2013, Milad Yousufi arrived in New York City for the first time. The eighteen-year-old pianist, composer and conductor had travelled with the Afghan Youth Orchestra, itself part of the Afghanistan National Institute of Music in Kabul, on a two-week tour of the eastern United States that included a concert at the Kennedy Center in Washington, DC and master classes at Boston's New England Conservatory. The crowning concert took place at New York City's Carnegie Hall and generated headlines such as '*Bolero* on Instruments Ravel Never Dreamed Of' (*New York Times*) and 'From a Land Where Music was Banned – to Carnegie Hall' (*NPR*). Milad would return each of the next two years to participate in summer camps, but in 2015 while he was in the United States, news came of worsening security conditions in Afghanistan, and he applied for political asylum. He had only the clothes and money he had brought with him for the summer camp. Happily, he would soon have a stroke of luck: while working at a fast-food restaurant as he waited for his application to be processed he met an immigration lawyer, Megan Jordi Brody, who helped him win a scholarship to study at New York City's Mannes School of Music. He jumped at the opportunity, but the keen pianist then discovered a new problem: where to find a piano on which to practice? Again, serendipitously, Jordi Brody was also a pianist, and, through a series of connections, she put Milad in touch with someone in New York who had a piano in need of some use. As it turned out, that person was Mary Kouyoumdjian. As Mary recalls, 'So this email was forwarded to me. And I thought, well, I have an upright piano. And I never practice it ... So I wrote and had tea with his immigration lawyer and with him' (interview with author, 2021). Over dinner, Mary, Milad and his lawyer got to know each other, and afterwards Mary asked if Milad would be willing to be interviewed as part of a compositional project. As she told us in an interview in 2021, this was both in recognition of the power of his personal story and as a way of strengthening his immigration case by demonstrating his contribution to the United States and integration into American life. A few days after they first met over tea, she and Milad had a lengthy interview and *Paper Pianos: 'You Are not a Kid'* was born.

As Milad says, this part of his life is 'like a movie' – in fact, a short documentary film has been made about his life (Water, Earth & Sky, 2021) and a television series has begun shooting – but as the main title of *Paper Pianos: 'You Are not a Kid'* suggests, the search for a piano that led Milad to meet Mary in New York City was not the first such search of his life. This is evident in the text of the piece, which opens with the image of the paper piano that gives the piece its overall title. Milad was born in Afghanistan in 1995 just before the Taliban gained control, forbidding music, art and other expressive media. Over the next six years, instruments were destroyed and musicians punished in accordance with the Taliban's strict interpretation of Islamic law. To compensate for the lack of a physical instrument, as a young child Milad drew a picture of the piano keyboard on paper and proceeded to 'play' in silence. It was not until shortly after the Taliban were ousted from power by the US-led invasion of Afghanistan in 2001, that Milad joined the newly formed Afghanistan National Institute of Music where he developed his pianistic and conducting skills. This led to a profusion of music and art making: when we speak during the summer of

2021, Milad's room in New York City is filled with paintings and sheets of music alongside multimedia pieces where colours swirl across staff paper. From the perspective of 2021, we can trace back a line of musical instruments that links the pianist-composer now making a way into the United States with the child in Afghanistan tracing the shape of music on sheets of paper.

Of course, this incongruous musical journey from paper representation to sounding instrument is only one portion of Milad's story. Although *Paper Pianos: 'You Are not a Kid'* begins with Milad's recounting of his precocious musical endeavours, he quickly makes it clear that childhood in Afghanistan was very different from childhood in the United States, to the point that its existence is not something to be grasped. The reminder that 'you are not a kid' was drilled into him by his father and further cemented by the indiscriminate violence he witnessed. In the interviews excerpted in *Paper Pianos: 'You Are not a Kid'*, Milad recalls the traumatic experience of losing childhood friends to outbreaks of violence and to the detritus of war, such as landmines. Others of his playmates were injured, maimed or orphaned, while still others disappeared. Childhood, in Milad's estimation, consisted of 'playing with bullets, hearing the rocket sounds, and witnessing my friend's funeral, and, I mean, seeing Taliban cutting the hands off the people and – I mean it was completely a violence there while I was living. That's why I'm telling, I never had a childhood'.[2]

Milad's voice lies at the heart of *Paper Pianos: 'You Are not a Kid'*, and as a listener it is easy to be captivated by its sound and textual content. Yet while the story the text recounts is crucial, it is not the sole component. In her composition, Kouyoumdjian combines excerpts from the recordings she made of their interviews with acoustic music for an ensemble of flute/piccolo, oboe, clarinet/bass clarinet, bassoon, horn, trumpet, bass trombone, two percussionists, piano, violin 1, violin 2/voice, viola/voice, cello, contrabass and conductor. Fittingly, the work opens quietly, with the recording of Milad's voice accompanied by what might be a percussion brush on paper or might simply be background noise. After a few moments, the piano is the first instrument to sound with sustained isolated tones hovering on the brink of melody, followed by short motifs in the strings and winds. Fragmentary allusions to folk music or to sirens snap into aural focus before dissolving again under pressure from

---

[2] Mary Kouyoumdjian, *Paper Pianos: 'You Are not a Kid'* (2016), 8′30″–9′00″. Recording available online: https://soundcloud.com/alarm-will-sound/you-are-not-a-kid-from-paper-pianos (accessed 24 September 2021).

throbbing, driving rhythms that build in bombast until panicked fluttering from the flute converges with wailing strings and straight-tone vocal slides. Out of this comes the sound of paper being crumpled, this time giving the eerie effect of the crackling of a fire. The sounds of the voice and the accompanying music continue to move in alternation, with sections of sparse instrumentation (often marked 'ghostly' or 'dreamy') allowing the vocal lines to come through clearly contrasting with sections where furious block chords from piano and percussion combine with anxious, buzzing trills and tremolos to fill in all available space. As Milad's spoken story turns to the tragic effects of children playing with bullets and mines, repetitive strings and electronics are accompanied by bell-like percussion to create a poignant backdrop to Milad's closing words about his lack of a childhood. The overall effect is of the sonic texture supporting, surrounding and contextualizing the voice; we might even say that it fills in the gaps left by the halting and incomplete nature of Milad's narration.

Paper Pianos: 'You Are not a Kid' was premiered in July 2016 by the New York–based contemporary chamber ensemble Alarm Will Sound in Columbia, Missouri, at the Mizzou International Composers Festival (the work's co-commissioners). A year later, Kouyoumdjian and Alarm Will Sound won a grant from the MAP Fund, an organization that supports 'new, original work by vocational artists and cultural practitioners that demonstrates a spirit of deep inquiry in form, content, or both, and that challenges inherited notions of social and cultural hierarchy' (MAP Fund 2021). The grant has allowed Kouyoumdjian and her collaborators to transform *Paper Pianos: 'You Are not a Kid'* into a large-scale multimedia work combining music and narration with hand-drawn animation by the Syrian artist Kevork Mourad and staging by Nigel Maister. Entitled simply *Paper Pianos* and scheduled to premiere in February 2023, this piece combines Milad's story with that of three other refugees and resettlement workers (Getachew Bashir from Ethiopia, Hani Ali from Somalia and Akil Aljaysh from Iraq) to 'explor[e] the dislocation, longing and optimism of refugees'.[3] Alongside new richness in collaboration, this transformation has brought with it a host of new concerns about how these stories should be told. As Kouyoumdjian relates:

---

[3] See the project description available online at: https://empac.rpi.edu/program/curatorial/residencies/2021/paper-pianos (accessed 4 January 2022).

When I interviewed them, this project was quite small and didn't have very big ambitions and didn't have any presenters attached to it or anything. And they had volunteered the interviews just because they work in resettlement, they want the story shared, and understood it would be in a musical context, and it would be staged, and it would be long, but I don't think [they] quite understood, as none of us did, that it would be presented by a large presenting organization.

So this next phase with them is having conversations between myself, presenters, and the interviewees about now that this project has grown – even though legally there's no requirement to compensate or do anything – what would feel good, like what could be useful, whether that is monetary, like a certain percentage of revenue goes towards them as individuals or their organizations, or if there's something we can be doing to really promote their organizations and their work in a way that is in line with their mission. ... you don't ever want to take advantage of something that was a very humble and modest hope at the beginning that has grown. (interview with author, 2021)

Kouyoumdjian's comments on the relational changes attendant on the transformation of *Paper Pianos* are instructive for any creative relationship – especially those based on storytelling – and we will return to them in a later section of this chapter. But they also point to a fundamental challenge for the sonic arts within social transformation efforts which is how to scale up projects without losing the kinds of personal connections that drove the art in the first place? When examining the story of *Paper Pianos: 'You Are not a Kid'* it is tempting to suggest that from the intimate drama of a search for a piano will come 'a meditation on the power of music, community, and communication to respond to the inhumanity of war, displacement, and violence' (Gordon 2019), but it is not immediately obvious what will be lost and what will be gained in the process.

Walking together

Although the circumstances of its inception are less serendipitous than that of *Paper Pianos: 'You Are not a Kid'*, *They Would Only Walk* has a similarly personal touch stemming from a connection between Kouyoumdjian and the arts education organization BSW's cofounder and Executive Director Yuki Numata Resnick. In 2016, Yuki was a violinist in Alarm Will Sound, and she played in the premiere of *Paper Pianos: 'You Are not a Kid'*. In our conversation in spring 2021, she remembers listening to Milad's voice on the spoken track and realizing:

First of all, I knew that I wanted to bring her to Buffalo so that our parents and the stories of our students could be told in their voice. ... I felt like, not only would Mary do our parents' stories, and our family's stories justice, but she would also do it in a way that would really bring in people from across the city – that the music wouldn't be polarizing. (interview with author, 2021)

Several years of conversations and fundraising were necessary before Kouyoumdjian could come to Buffalo as BSW's inaugural Composer-in-Residence, and *They Would Only Walk* was commissioned in 2019.

Beyond this connection, digging a little deeper into BSW's own story reveals a similar combination of the hard work, musical talent and courage Milad demonstrates. BSW began in 2014 as a response to an outreach performance undertaken by a group of musicians at a primary school in Buffalo, New York. In one interview, Yuki recalls that they had played the slow movement from Brahms's Piano Quartet:

> We asked the students what they thought of it, and how did the music make them feel. And you know, and there was a little boy, and he was under a desk, and he looked around seven years old. And he said, 'You know, it sounds like I love you.' And so I think it was that response to the music that suddenly made us stop and realize that the music was affecting the students in the room in a very different way than we were previously used to.

The school Yuki mentions, Buffalo's PS 45 International, is on the West Side of Buffalo in the heart of the city's refugee and immigrant community. Due to funding pressures in Buffalo's public schools and the area's socioeconomic makeup, many of the students have limited exposure to instrumental music instruction, so Yuki and fellow cofounders Elise Alaimo and Virginia Barron started BSW as a non-profit strings education programme. As of 2021, they serve 99 students and have recently opened a second location serving the Hispanic and Latinx community.[4] Beyond providing a high-quality musical education, they 'strive to foster greater understanding and compassion across Buffalo, NY, offering a sense of belonging to refugees, immigrants, and those who have long called Buffalo home. ... by lifting up the voices of our students and parents, we cultivate youth to be agents of social change' (Buffalo String Works 2021a).

The educational mission of BSW means that from the outset, Kouoyumdjian's residency was designed to be 'as immersive of an experience as possible': something

---

[4] Yuki Numata Resnick, interview with author, 2021.

that involved a two-way transfer of knowledge, not a composition that would take place at a distance or as the product of isolated trips to record interviews. Thus, on the composer's trips to Buffalo during the autumn of 2019, much of her time was dedicated to working directly with students in workshops where students 'compose[d] in real time' through experimenting with unusual sounds and techniques. As Kouyoumdjian recounts:

> One day ... we talked a lot about methods of transportation, because everyone took a method of transportation to come to the United States. And we were very literal about it, like okay, so imagine you're on a plane: how can you make your instrument sound like the engine? Or like, what does it feel like to be up in the air and they would all do their own extended techniques to do that. (interview with author, 2021)

As this description demonstrates, these workshops were not only enjoyable and educational but also the sounds they generated fed directly into the piece and offered an opportunity for 'every child ... to feel invested in the creation of the work' rather than only those who were interviewed. This concern for broad involvement extended to the construction of the music itself, which Kouyoumdjian describes as the result of 'writing with children' rather than for them. One consequence of this approach is that there is a core string quintet that plays the short, vigorous melodies and more complex rhythmic outbursts throughout the piece, while the other strings provide a rich textural backdrop though rhythmic ostinati. Overall, the music is consonant and unobtrusive in its cradling of the text. *They Would Only Walk* is thus positioned as at once suitable for young performers and intermediate players yet not (merely) a children's piece: something 'that they can be proud of, and want to listen to, and can be performed. And even as they grow older, that they can perform' (interview with author, 2021).

In addition to working directly with BSW's young musicians and other musical groups, Kouyoumdjian also spent time getting to know the students and their families during her time in Buffalo. She interviewed a number of those involved in BSW about their own or their family's experiences of coming to the United States, and the voices of four students (Mohamed, Hung Kee, Yamilette and Salina) as well as those of five family members and one interpreter appear in the final piece.[5] In keeping with the population that BSW serves, the interviewees

---

[5] The family members are listed in the programme as Mariana Gonzalez, Myintcho Lay, Dim Lan Lun, Eh Tah Mu and Chai Sen, along with Steven Sanyu, who served as Chai Sen's interpreter.

trace their heritage around the world: four of the adults come from Burma and another was born in Venezuela, while the children include one born in Morocco, one in Malaysia, one in Eritrea and one, the daughter of a migrant from Mexico, born in Buffalo. Fragments of these stories are interwoven within the piece so that descriptions of a pre-conflict childhood in rural Burma are juxtaposed with childhood in a Venezuelan city or displaced into a Thai refugee camp.

In the central section, themes of travel come to the fore whether that means walking through the forest and traversing rivers at night or boarding a bus to Bangkok followed by airplane journeys to South Korea and New York. Here, too, we catch the words that give the piece its title in a section where Yamilette describes her father's arduous journey on foot across the United States' southern border. He and his companions 'would only walk at night', we discover, and hide during the day for fear of encountering the Border Patrol.[6] The final section of recorded speech turns to life in Buffalo and, for both students and their families, hopes for the future. Parents speak of the desires for their children to become good people, or to become musicians or simply to share what they have learned with the next generation. Yamilette and Hung Kee describe their desires to help others, with Yamilette explaining that a trip to Mexico has led her to want to become a doctor so that she can help children there. The text of the piece closes with two of the adult speakers in a quasi-conversation – almost finishing each other's sentences – about how music can provide happiness, healing and a sense of bonding that brings families together.

The nonlinear nature of the storytelling and the presentation of small fragments of text in different voices can make it difficult to piece together a complete or coherent story in *They Would Only Walk*. For example, on our first listening, we were initially unsure as to how many different people were speaking or whose storyline we had heard previously. Even though the accompanying programme notes identify the speakers as a group and present a transcript of the text, this aural uncertainty is likely to have been a familiar experience among others of BSW's initial audience. Ultimately, however, it is the moments these recordings capture along with the ineluctable sense of being addressed that generate the greatest impact.

Kouyoumdjian finished the score at the beginning of March 2020 ahead of *They Would Only Walk*'s scheduled premiere that June, but as the Covid-19 pandemic developed, it became clear that the traditional live premiere would be impossible.

---

[6] *They Would Only Walk*, 7'30"–8'14".

Undaunted, the format of the premiere was re-imagined, and in the autumn of 2020, students and members of the Buffalo Chamber Players came in one by one or in small groups to record their contributions in Buffalo's Asbury Hall in both audio and visual format. This complete, the filmed performance of *They Would Only Walk* would become the centrepiece of a virtual event: the seventh Annual Benefit Concert for Buffalo String Works, held on 20 March 2021 and available to stream online for the following week. In addition to *They Would Only Walk*, the concert featured recorded performances by a number of well-known musicians from Buffalo and beyond, including Aaron Dessner, Matt Berninger, Joshua Roman, Ben Lanz, Lisa Hannigan, Lauren Sprenglemeyer and Richard Reed Perry, Time for Three, Potter's Field and Gail Ann Dorsey (see Buffalo String Works 2021a). Other familiar components of an in-person benefit concert, including sponsorship by local companies and a silent auction, were similarly adapted to the virtual format.

From a research perspective, the shift in presentation meant that we were able to join the virtual audience for the benefit concert in March 2021; we also took advantage of the event's streaming format to listen multiple times to the pieces over the following week, making ample use of the resulting ability to control the musical-temporal experience. This results in a markedly unusual and iterative experience of a premiere of so-called classical music (though one familiar from music videos and films), but – even more unusually – several aspects of this experience were shared with the performers themselves. One BSW student told us that her family had watched the concert on YouTube at home while she had watched from somewhere else.[7] The digital format of the performance was further enhanced by the use of BSW's YouTube channel to provide a rich context for the work – for instance, by hosting several brief videos that introduce BSW students who feature in the piece alongside brief commentary videos on *They Would Only Walk* from Kouyoumdjian and one of BSW's instructors, Shannon Reilly (Buffalo String Works 2021b).

Honouring the stories: Ethnography, ethics and composition

Taken together, *Paper Pianos: 'You Are not a Kid'* and *They Would Only Walk* offer insight into a process I call here *ethnographic composition*. Ethnographic

---

[7] Yamilette García, interview with author, 2021.

composition combines principles of ethnography within institutional and creative structures familiar in Western art music. For example, Kouyoumdjian's extended engagement with BSW students and their families in Buffalo and with Milad constitute a kind of fieldwork in which the composer's participation in aspects of daily life leads to deeper relationships. In addition to this general principle of active involvement, Kouyoumdjian gathers data and compositional material via open-ended interviews that are equally engaged in the informational and the phenomenological aspects of individual lives. Even though the act of composition itself takes place after the interviews (analogous to the ways in which field notes and transcriptions feed into a finished ethnography), it might be thought of as a form of writing that takes place in conversation with those people whose stories are crucial to it.

In other ways, Kouyoumdjian's compositions are recognizably influenced by the institutions and creative practice of Western art music. For example, both of the pieces discussed here were commissioned by arts ensembles or organizations and premiered as parts of recognizable concerts. The evening-length multimedia work *Paper Pianos* forms part of Kouyoumdjian's compositional portfolio for her doctorate from Columbia University. On a still-more fundamental level, although she is sceptical of the title 'composer', Kouyoumdjian is recognized nonetheless as the composer of each work and retains a position of creative agency in relationship to them.

Thus, even though Kouyoumdjian's works are well within a recognizable mainstream of contemporary Western art music in terms of their sonic content and in some aspects of their institutional identities, the ways in which they are put together reveal a compositional attitude distinct from, on the one hand, an approach that sees a composer's job as setting a preexisting text within a musical narrative, and, on the other hand, from one that treats prerecorded voices primarily as a sonic resource. The impact of this is more evident when we consider Kouyoumdjian in relation to other contemporary composers who have worked with prerecorded voices. One particularly influential composer is Steve Reich (b. 1936), who began working with recorded voices on tape in the 1960s. Early Reich tape works include *It's Gonna Rain* (1965) and *Come Out* (1966), both of which loop short vocal recordings that fall out of sync over time, thereby rendering the original words unintelligible. As Reich comments, working with speech in this way meant one could 'keep the original emotional power that speech has while intensifying its melody *and* meaning through repetition and rhythm' (Reich 2002: 20).

Different Trains, Reich's 1988 composition for string quartet and tape, takes a different approach in using extracts from recorded interviews with Holocaust survivors and others to build up an aural image of the divergent histories facing Jews in America and in Europe during the mid-twentieth century. Reich suggests that his embedding of these voices into the string quartet's musical texture via what he terms 'speech-melody' together with the inclusion of recorded fragments means that *Different Trains* 'presents both a documentary and a musical reality'. It thus 'begins a new musical direction. It is a direction that I expect will lead to a new kind of documentary music video theatre in the not too distant future' (Reich 1988). Reich's advocacy of speech and the voice as a repository of reality – and music's contribution to bringing that reality into being – has inspired numerous composers to take up a similarly documentarian role. Yet, if Reich's music is documentary, it is a documentary style in which the composer's ear and hand play a significant, unacknowledged and potentially distorting role in selecting and arranging the voices. In her analysis of *Different Trains*, the accumulation of misheard and re-arranged phrases lead Amy Lynn Wlodarski to characterize the piece as ultimately 'Reich's own Holocaust testimony, one crafted from the voices of witnesses other than himself' (2010: 104). Moreover, in the case of *Its Gonna Rain* and *Come Out*, the technical approach to composition entailed dissolving some key aspects of those original voices and their experiences until what remains is only rhythm; given that those original voices belong to Black American men in the 1960s, contemporary scholars such as Sumanth Gopinath (2009) have suggested that Reich's treatment of these voices as sources of sound is a form of violence with disturbing racial undertones. In each of these cases, there is no opportunity for the vocal subjects to answer back and no clear indication that the composer saw himself in a reciprocal relationship with the individuals whose voices he manipulated. These complications and what Kouyoumdjian (2021: 45) terms Reich's 'delayed accountability' for them, demonstrate the need for a sensitive treatment of individuals and the sounds of their voices based on a firm ethical understanding of the relationships contingent on the compositional process.

I focus on Reich here, rather than other on composers who have made extensive use of oral recordings such as Gavin Bryars, because Kouyoumdjian has explicitly cited Reich's work, and particularly *Different Trains*, as crucial to her own development: *Bombs of Beirut*, which was inspired in part by *Different Trains* and was written for the same ensemble (the Kronos Quartet), remains one of her formative works. Reich thus serves both as an early model for

Kouyoumdjian's documentary musical style and as a warning of the difficult ethical issues surrounding working with recorded voices and personal narratives (2021: 45–6). In navigating these situations, Kouyoumdjian has developed a distinctive compositional ethos that foregrounds the shared human connection forged between the interviewer and interviewee, or between the speaker in a piece of music and the audience to whom they speak. In this, she is distinguished from other composers who have generally relied on preexisting footage. As she says:

> I do think that what I never want is to feel like the people that I'm interviewing are subjects – almost like a scientific approach. Because if I'm speaking with them, oftentimes, their experiences are so close to my family's, and I really connect with that and know how much it takes to share and [I] want to make sure that there's some human – if I'm asking for human connection between an audience member and their recorded testimonies, I have to have a human connection with them as well. (interview with author, 2021)

As this quotation suggests, Kouyoumdjian experiences demands of intimacy and trust as a result of her personal identification with her interlocutors even as her desire to share their stories publicly entails a certain amount of journalistic probing of their experiences. This ethos influences not only the semantic content of her work but also the practical ways in which she enfolds the idiosyncratic timbres, accents and rhythms of the interviewees' speech within instrumental music that situates itself in dialogue with the text.

By prioritizing what she terms 'human connections', Kouyoumdjian is making an ethical choice that shapes her interactions and one that echoes in the distinction that can be drawn between storytellers and story takers. As the Italian feminist philosopher Adriana Cavarero has argued, most modern knowledge comes to us via a story-taker – 'the one who solicits and listens to life-stories told by others, in order to then transcribe them into the scientific canons of his discipline' (2000: 64). Cavarero links the story-taker to the archetypal modern figure of the psychoanalyst, but the anthropologist, oral historian or composer-documentarian might furnish equally salient examples. The problem, as Cavarero sees it, is that the story-taker may not return that story to its originator, resulting in a broken or incomplete relationship between the two individuals.[8] This

---

[8] This image of story-taking as a kind of anthropological extractive industry resonates strongly with contemporary concerns over power relations in ethnography.

'asymmetrical and unbalanced dependency' (Cavarero et al. 2018: 86) between individuals raises questions about voice and the ethics of storytelling which are themselves founded in a conception of the self that sees the individual not as narrated but rather as narrat*able*. For Cavarero, this seemingly minor linguistic shift between the narrated and the narratable self conceals a more fundamental proposition, which is that in being narratable – in being reliant on the narration of others for its completeness – the self is revealed as fundamentally relational and inclined towards the Other. Only by acknowledging the extent to which each self relies on another to furnish its story can such relational asymmetries be adequately addressed.

Although Cavarero is a philosopher, and not a musicologist, her philosophical thinking is intimately connected to the social and sonic practices of storytelling which anchor her work on subjectivity and the voice. One suggestive possibility for the application of Cavarero's principles within music is found in her book *For More than One Voice* (2005), where she extends her concern with interpersonal relationships into a philosophy of vocal expression. Here she offers a counter-history to the philosophical tide that has privileged the semantic over the sonic and that has treated the voice as (merely) a stand-in for the agency of the individual subject. Cavarero's displacement of traditional Western philosophical insistence on subjectivity in favour of intersubjectivity is also a rejection of Jacques Derrida's (1976) conception of the logocentric voice as a soliloquy experienced primarily (or only) by the speaker.[9] To have a voice is always to talk with another. To imagine otherwise is to fail to recognize the reality of individual interwovenness with others and thereby participate in a process of atomization that strives to relieve us of our responsibility to listen and respond.

Extending Cavarero's vision of the self as constructed through relations of telling and retelling of stories to a musical context, we begin to see the potential for a distinctively musical storytelling to transform the intersubjective relations between storyteller, composer and audience. If the propensity to insist on the self as narrated – and especially on the narrated self as contained in and by written language – is to transform a being into a book or a who into a what, to return a self-narrative to its originator via a musical 'voicing' is to acknowledge the uniqueness of that individual and his or her life story. When

---

[9] In other words, it is not Derrida's critique of logocentrism or the interest in the materiality of speech that Cavarero wishes to overturn, but rather Derrida's neglect of the relational aspects of voice.

this occurs, it is possible for the composer/story-taker to become a storyteller in a mutual relationship of gifting and re-gifting of stories sounded through many voices.

The question of who speaks, for and to whom are ever-present in Kouyoumdjian's ethnographic compositions, and an attentive reading of her comments reveals an evolution in her own thinking on the matter of voice and its relationship to agency within music. For example, although she has previously discussed her work in terms of offering a voice to others, now when she reflects on the process of composing with other peoples' stories, she rejects that idea. Quoting the documentarian and scholar of films Michael Rabinger, she writes in her doctoral thesis, 'Speaking on behalf of others is almost a disease among documentarians … Behalfers make it their business to represent those without a voice, which in the end is everyone unable to make films themselves' (Rabinger 2009: 357; quoted in Kouyoumdjian 2021: 46). The literature on composers and those whom they depict in their music is less well developed than in film documentary, but a similar danger of representation and misrepresentation underpins the idea of a musical 'behalfer'. A careful examination of Kouyoumdjian's compositional process suggests that she does not disavow the privileged position of composer nor its component specialist technical knowledge and creativity, instead, she uses these abilities to counter (mis)representation in two ways. First, as mentioned above in relation to *Paper Pianos: 'You Are not a Kid'*, she works closely with individuals who want to tell their story in order to produce a work that engages with the stories as those individuals want to tell them. Second, as demonstrated in both her workshopping with BSW and in her mentorship of Milad, she strives to disseminate those same technical skills to better enable those original storytellers to participate in their stories' musical incarnations. Unlike Reich, for whom the sounds of recorded voices were often a way of generating musical melodies through the distortion or dissolution of the original, Kouyoumdjian uses effects such as compression, echo and reverb sparingly to preserve both a strong sense of coherent narrative and the sonic integrity of the voice itself. Unlike the documentarian mode which speaks on behalf of others to a wider audience, Kouyoumdjian sets these stories into music in order that they may first be given back to their tellers. By prioritizing the needs of the individuals with whom she works as part of the compositional process, Kouyoumdjian puts into practice some key ethical principles of musical storytelling.

Remediating the past

Paying attention to the relational aspects of musical storytelling clarifies its ethically charged foundations, but Cavarero's framework is based on a model of private interpersonal speeches between individuals who know one another. In order to think about relationality on a larger scale, we need to consider how these stories are transmitted – alternatively, remediated – beyond that initial relationship. Remediation, or the repeated translation of a concept or process from one format to another, is a familiar concept from media studies. There it is most commonly discussed in relation to specific convergences of media or their material properties, such as the possibilities for digital media to adapt and transform preexisting visual, sonic or literary media forms. From this perspective, this chapter's adaptation of ethnography to describe the creative processes of Kouyoumdjian's compositions might be considered an example of remediation occurring between the disciplinary procedures of music studies and those of oral history and anthropology, while Kouyoumdjian's own discussion of musical documentary indicates a cross-fertilization between the filmic and sonic arts. Yet remediation is not simply another word for interdisciplinarity; it possesses what Bolter and Grusin (1999: 5) call a 'double logic' in which the recycling of media and material simultaneously draws attention to the process of mediation and the presence of a given medium even as it may also burnish that medium's claim to transparency. Within this double logic of hypermediation and immediacy, Kouyoumdjian's *re*mediation of audio recordings within her compositions suffuses her work with the impression that it captures reality even as it highlights the multiply-mediated nature of what we hear. For example, Milad's discussion of painting paper pianos as a child is accompanied by a brushing of paper whose gentle rhythms and timbre call up images of the act of painting Milad has just described. As this continues, it sweeps the listener into Milad's imagined music where rippling piano arpeggios suggest the young composer's imaginary engagement with a real instrument.[10] These sounds of paper and of the piano suspend the listener between the audiovisual present (i.e. a percussionist and pianist present on the stage) and the multiple pasts invoked, first by the sound of Milad as a young man and then by the textual description of

---

[10] Mary Kouyoumdjian, *Paper Pianos*: 'You Are not a Kid' (2016), 1'50"–2'30". Recording available online: https://soundcloud.com/alarm-will-sound/you-are-not-a-kid-from-paper-pianos (accessed 24 September 2021).

his childhood. This transfer between different compositional elements requires an attentive and imaginative form of listening to both the narrative and the sonic contours of Milad's voice.

Looking beyond these two works, two further aspects of remediation are significant in the context of the sonic arts and conflict transformation: specifically, remediation indicates the reiteration that lays the foundations for collective understandings of the past and present, and it also contains within it the etymological resonance of 'remedy' (see Erll and Rigney 2009; McMurray 2021). The first of these suggests that any given process of remediation is a driving force for further remediations that coalesce into cultural memory; the second highlights remediation as a kind of return, a making-up-for or a giving back. In the case of music, the reuse of other sonic material re-contextualizes those sounds, but the reiterative nature of musical performance also encourages repeated listenings that confirm these new contexts. As I have argued elsewhere (Phillips-Hutton 2018), the use of archival fragments to create new musical narratives can be a crucial step in generating more inclusive and multifaceted histories. When seen in conjunction with Cavarero's concept of relating narratives, this suggests that a specifically musical remediation is not simply a question of tracing the intermedial transformations of musical sounds but of paying attention to the ways in which participants in these musical encounters use sound to structure sociality as a potential step towards remedying damaged and broken relationships.

Relational aesthetics

Thus far, this chapter has focused on the implications of compositional processes and the consequent relationships between Kouyoumdjian and her interlocutors. But in addition to these individual ethnographic encounters in the preparation of a piece, ethnographic composition involves an eventual audience as a third party. Moreover, these pieces are intended not solely as aesthetic experiences, but also as impetus for human connection, empathy and ultimately social change. As Kouyoumdjian explains in our interview, 'My most humble hope in all of this is that there's just a spark of empathy between the listener and the interviewee … And hopefully, then empathy translates to cultivating a culture that is more ready and willing to act on direct change.' This compositional intention leads to a consideration of what a music that engages deeply with the relationships that flow from a polyphony of voices might offer in terms of the remediation of both sound and relationship.

The stories these compositions contain are both compelling and compellingly told, but the successful communication of narrative content is only one contributing factor to their significance. When it comes to bringing about positive social outcomes, the textual content may ultimately be of less importance than the means by which it is communicated, which is to say that the characteristics of music as a distinctive sonic and performed art plays a crucial role. In arguing that *They Would Only Walk* and *Paper Pianos: 'You Are not a Kid'* are ultimately performances in which the aural characteristics of a voice or an instrument and the relationships they encourage can be as significant as the words that the piece sets, I am drawing on the work of Nina Sun Eidsheim (2015), who discusses sound as a Geertzian 'thick event', or a complex amalgam of phenomenon, material surroundings, perception and reception, in which the techniques of listening we employ become a part of the sound rather than a response to it. However, I am also arguing for the importance of listening and acknowledgement as a reciprocal act in an effort to use storytelling as a means of encouraging resilience and reconciliation. In particular, music's relational aesthetics, its storytelling capacity and its potential to model sociality are all significant factors in music's potential for social transformation.

While numerous studies (Colvin 2018; Davis and Meretoja 2017; Emberley 2014; Myers 2016) have argued that the opportunity to share personal experiences and to have those acknowledged by the wider world is a crucial part of healing for individuals who have experienced traumas (e.g. those engendered by forced migration and violence) – and recent years have witnessed a surge of literature dedicated to teaching reconciliation through storytelling (see *inter alia* Beckerman and Zembylas 2012; Bell 2019; Ng-A-Fook and Llewellyn 2019) – it is less evident that the specifics of a given narrative are significant for the audiences of those stories. In her critique of the implicit power relationships present in many storytelling efforts, Serene J. Khader asserts that 'it would be a mistake to take the call of victims' stories as possessing any determinate content – even for empathetic listeners' (2018: 21). Following Khader's analysis, the impact of Kouyoumdjian's work must come from something beyond the textual content – even when that content is compelling. A closer look at the mechanisms by which the storytelling in *They Would Only Walk* and *Paper Pianos: 'You Are not a Kid'* can contribute to positive social outcomes highlights the crucial importance of relational aesthetics in music; in other words, how this music creates and sustains peculiar kinds of relationships among its participants.

Contemporary academic interests in relational aesthetics hail from many different quarters (e.g. art theory and criticism, popular music studies, and anthropology) but all centre on a core question: how do aesthetic objects generate intersubjective relationships with the world? When applied to music, thinking about relational aspects foregrounds the ways in which sounds are organized in relation to one another in space and time as well as the ways in which music making generates distinctive arrangements of human bodies. On both material and metaphorical levels, the relationships music creates are at the centre of the performances of *Paper Pianos: 'You Are not a Kid'* and *They Would Only Walk*. For example, in *They Would Only Walk*, Kouyoumdjian juxtaposes different instrumental sounds, voices and stories in order to generate its overall musical structure. The resulting dialogue between instrumentalists and recorded speakers creates new perceptual relations among the different sounded layers. At the same time, those sounds reveal intergenerational relationships between parents and children and interspatial relationships between past homes and present ones. As Yamilette mused about her contribution to *They Would Only Walk*:

> 'They Would Only Walk': I would make that part of my story – well, my dad's story. Because what he did was extremely incredible when he was 17 or 18. That's when he crossed the desert to live the American Dream. And I feel like the title also has meaning – like a *lot* of meaning. And there's obviously some meaning that I'm not aware of because I'm still young, I'm still learning. So I feel like hopefully in the future I will come back to the music and the title to see what my thinking was before and after. (interview with author, 2021)

Most significantly, these works call their audiences into an acoustically structured relationship – to listen with or alongside one another as well as to the narrators. These multiple acts of listening are not merely a response to the sounds of the voice or of the music, but actually they constitute a part of the performance as a 'thick' sonic event in which music making is understood as an art of relationship.

Effects

Evaluating Kouyoumdjian's work in *Paper Pianos: 'You Are not a Kid'* and *They Would Only Walk* requires balancing several different metrics, including the works' efficacy for the refugee storytellers, for the musicians and Kouyoumdjian, and for audiences, as well as their success as aesthetic statements. While each of these provides different insights, in this analysis I focus on the reactions of those most intimately involved in the work – namely, the interviewees.

Milad is effusive in his description of Kouyoumdjian's influence on his life, but when he turns to his involvement in *Paper Pianos*: 'You Are not a Kid', he is direct about his motivation: 'I just wanted to tell my story to the world and [to show] the bigger picture, but to [make] people here in New York and Denmark to be more appreciative of what they have, because sometimes I find them like take this, take things for granted here' (interview with author, 2021). When Yamilette was asked about working with Kouyoumdjian, the self-described 'NOT a people person' enthused about her experience (interview with author, 2021). At first, she had mixed emotions as excitement over the opportunity to work with a composer and nervousness over what might go wrong vied with pride in being a part of BSW and in sharing her father's story, but in the end the sense of pride together with the feeling of safety and trust fostered by Kouyoumdjian won the day and she rose to the challenge. She even said she would be willing to do it again!

Due to the Covid-19 restrictions on live music making in groups, the first opportunity for participants in *They Would Only Walk* to hear the entire work took place after BSW lessons. Yamilette recalled:

> I was excited, but nervous because Miss Mary put in some parts, not the whole story, because obviously, it's not long enough. So I was kind of nervous on what parts she would put into it. But I thought it was exciting to be my voice heard to be a part of the piece, like with other people to watch, [so that] other people can know. (interview with author, 2021)

Yamilette's sense of mixed emotions extended throughout those involved in *They Would Only Walk*. The initial listening session was limited to those whose voices (or whose parents' voices) appeared in the piece, both to limit any possibility of embarrassment and to provide a welcoming space. As Yuki related, many were moved:

> There was also a Venezuelan parent who spoke and she said, she was like, wow, you know, it was really wonderful to hear these shared stories and to kind of understand each other's stories, even though their lives and where they grew up are so different – but there were these shared moments.
>
> it was very emotional for them too – especially our Burmese parents, because they were hearing themselves talk about leaving this situation that had been improved. And now they're seeing their country, and they're seeing their friends

and family who are still there, going through exactly what they talked about. (interview with author, 2021)

As this suggests, the juxtaposition of positive social connections between refugees with different life stories together with the very real anguish experienced by those who watch ongoing conflict from afar is at once confirmation of the possibilities for musical storytelling to generate relationships and a sobering reminder of music's limitations.

At the end of our interview, Yamilette asked me what inspired the project to write about how composers and musicians work together. This provided a welcome opportunity to articulate the origins of this research in a belief that the stories we tell each other about who we are and where we come from are important both personally and socially. Music can help us tell those stories, but it can also help us collectively connect with the stories told by others. In the end, as Yuki said in our interview:

> I think it was our hope that people would see and hear this, and it would just put a face – it would put a face and a name and a story to people, it wouldn't just be like, 'Oh, the refugees, the immigrants, the Burmese people,' it would actually be like, 'Oh, actually, I might better understand what that means.'

***

The premiere of *They Would Only Walk* took place six weeks after the military overthrow of Burma's elected government in February 2021 and the resulting – and ongoing – violent suppression of protest and dissent. The situation in areas under the control of the Tatmadaw and in areas contested by militias, not to mention among those displaced into the forest or refugee camps, remains at once murky and dire. The first draft of this chapter was written as the Taliban swept across Afghanistan and returned to power in the summer of 2021. When I spoke to Milad that summer, he had yet to receive an answer to his asylum request, despite living for six years in the United States. He was increasingly worried about the fate of friends left behind. Likewise, the future of the Afghanistan National Institute of Music with whom he made his first trip to the United States is uncertain; its founder and director Ahmad Sarmast was severely injured in a bombing at a concert in 2014 after which the Taliban accused him of corrupting Afghan youth. Although it is too early to tell what the outcome will be in either of these situations, these are sobering and (at least momentarily) highly visible reminders of the violent realities faced by individuals and communities around the world.

Yet the news does not – cannot – give us the full story. The stories told in *Paper Pianos: 'You Are not a Kid'* and *They Would Only Walk* exist in a counterpoint to the news stories that splash across the headlines and then eventually fade. These pieces do not purport to capture the full drama of world events, but that should not cause us to overlook their ambitions. These pieces combine electronics, acoustic sound and voices into musical events in ways that provide space, time and a microphone for individuals and communities that might otherwise struggle to be heard. But beyond the appeal of simply 'having a voice' (understood as a heuristic for power and identity), these pieces should point us towards thinking not about the composer as granting the speaker their capacity to speak but rather about the relationships that speech and story are always creating. In these instances, what is most significant is not the technical process that underpins these musical stories but rather what happens in the rooms where these stories are told, in the moment the musicians bring sound forth, in the instance of being called on to speak and to listen.

5

From noises of conflict to dissonant sounds of reconciliation in the Northern Irish theatre

Stefanie Lehner

Introduction

In Kabosh's production of *Those You Pass on the Street* (McKeown 2014), the tinkling sound of a doorbell interrupts the trifling conversation between Elizabeth Farrell (played by Laura Hughes) and Ann, her sister-in-law (played by Carol Moore), announcing an unexpected visitor – Frank (played by Paul Kennedy), who has chased up Elizabeth's complaint about unsocial behaviour outside her house. As Elizabeth's husband, Michael, who worked as an Royal Ulster Constabulary (RUC)[1] Superintendent at Crumlin Goal was killed by the Irish Republican Army (IRA), Ann is visibly shocked when she learns that he works for Sinn Féin and questions Elizabeth's decision to go to them as the local 'elected representatives here' (McKeown 2014: 10). This discovery almost leads to a quarrel between them. Comparable sonic interjections arise in Kabosh's *Green and Blue* (McKeown 2016), which employs two familiar sounds of warfare that suddenly suspend the conversation between the RUC Constable and the An Garda Síochána[2] officer stationed on their respective sides of the (Northern) Irish border. Firstly, we hear a series of overlaid sounds creating the effect of a bomb explosion, reverberating within the landscape. Secondly, we are confronted with the noise of a rifle shot resonating across an open space, suggesting a sniper at the border. Akin to the doorbell in *Those You Pass*, both

---

[1] The RUC was the former police force of Northern Ireland from 1922 to 2001. It was superseded by the Police Service of Northern Ireland (PSNI) in 2001. It was renamed and reformed as a result of a fundamental policing review, the Patten Commission, as part of the 1998 Good Friday or Belfast Agreement.

[2] An Garda Síochána is the police force of the Irish Republic (commonly also referred to as Garda or Guards).

noises suddenly interrupt the scene, intruding the aural space of the theatre in a notable way.

What is the effect of these specific sounds? How are audiences affected by them? In *Theatre Sound*, the sound designer John Leonard argues that sound effects are usually employed in theatre performances for five reasons: (1) to provide the audience with information about the specific context of the play, (2) to illustrate a specific textual reference that the character or action in the play is referring or reacting to, (3) to create a specific mood or atmosphere, (4) to provide an emotional stimulus for the audience, and (5) to act as cues to reinforce stage action (Leonard 2001: 142). While the doorbell in *Those You Pass* certainly works as a cue, and the explosion and gunshot in *Green and Blue* provide an emotional stimulus, they are also much more than that. Following Katharina Rost's (2011) analysis of the powerful impact of intrusive noises in theatre performances, this chapter suggests that these sounds notably affect both onstage characters and offstage audiences, penetrating their physical and emotional sphere in a radical way. In looking at four Northern Irish performances by three theatre companies that were project partners on this project, this chapter proposes that intrusive noises in these plays are associated with conflict, which can attain a transformative power that enables new perspectives and meaning making to emerge, which, in turn, can open new pathways to approach reconciliation.

The four plays considered in this chapter can be considered as Post-Agreement theatre, conceived, written and produced after the 1998 Belfast or Good Friday Agreement, which sought to put an end to the three-and-a-half-decades of war and violence, euphemistically known as 'The Troubles', by setting up a nationalist and unionist power-sharing government in Northern Ireland as a basis for lasting peace. Yet the consociational model that underpins the devolved assembly has been criticized for naturalizing rather than transcending the divisions of the two dominant ideological blocs, which threatens to occlude a vast range of other relations, positions and concerns (Shirlow 2004: 196). Hence, although the Agreement envisions a 'fresh start' and promises a 'new beginning' (Agreement 1998: §2), the prefix does not stand as a temporal marker that delineates a distinct break from the past: instead, these plays deal with not only the unresolved issues of the legacy of Northern Ireland's past but also new and ongoing challenges that haunt the peace process. As such, they can be considered to demarcate the 'liminal coordinates' of a still troubling past and a precarious 'fresh' future, signalling what Birte Heidemann (2016: 8) identifies as a sense of 'negative liminality', which she describes as 'a disabling condition which, in the context of Northern Irish literature, pertains to a suspended state of (fictional) subject positions that resist closure and resolution.' But instead of considering this liminality as a restrictive state, this chapter draws on the theories of the theatre critic Erika Fischer-Lichte to reclaim the transformative and emancipatory potential of liminality by exploring the specific affective impact that sounds in these four productions have on their audiences.

The first part of this chapter will analyse the sound effects of the two above-named Kabosh productions: *Those You Pass on the Street*, first produced by Kabosh Theatre Company in 2014, and *Green and Blue*, which premiered in 2016. Both are written by Laurence McKeown and directed by Paula McFetridge. Both plays address the complexities of dealing with the legacy of the Northern Irish conflict and toured extensively within the UK and Ireland as well as abroad.[3] *Those You Pass on the Street* was commissioned by Healing Through Remembering, an independent organization dedicated to 'working on how to deal with the legacy of the past as it relates to the conflict in and about Northern Ireland' (Healing through Remembering Website).[4] *Green and Blue* takes its name from the 'Green and Blue: Across the Thin Line Project' and is inspired by the stories collected as 'Voices from the Vault', an oral history archive of former personnel serving the RUC and An Garda Síochána officers, who recorded their experiences as police officers during the Irish conflict.[5] The second part of the chapter shifts its attention to two more recent productions: Tinderbox's 2019 adaption of Alfred Jarry's *Ubu Roi* by artistic director Patrick J. O'Reilly, *Ubu the King*, and TheatreofplucK's 2018 performative verbatim audio walk *So I Can Breathe This Air*, written by Shannon Yee and based on interviews with The Rainbow Project's Gay Ethnic Group (GEG), which was originally titled *Multiple Journeys (of Belonging)* (2017). This analysis draws on Stefanie Lehner's affective experiences of these performances as well as participant observation, post-show discussions and questionnaires and, in the case of the Kabosh plays, Image-Theatre workshops.

---

[3] For more details, see Kabosh's website. For *Those You Pass*, see online: https://kabosh.net/production/those-you-pass-on-the-streets/. For *Green and Blue*, see online: https://kabosh.net/production/greenandblue/ (accessed 18 October 2021).

[4] Available online: http://healingthroughremembering.org/who-we-are/history/ (accessed 18 October 2021).

[5] Available online: https://www.green-and-blue.org/voices-from-the-vault/ (accessed 18 October 2021).

Part 1: Kabosh's *Those You Pass on the Street* (2014) and *Green and Blue* (2016)

Agitatory noises of conflict

In *Theatre Noise: The Sound of Performance*, Lynne Kendrick and David Roesner (2011: xv) suggest that 'theatre provides a unique habitat for noise'. They explain, 'It is a place where friction can be thematized, explored playfully, even indulged in: friction between signal and receiver, between sound and meaning, between eye and ear, between silence and utterance, between hearing and listening.' As an 'agitatory' acoustic aesthetic, it seeks to make us alert and asks us to reflect on the potential meaning of this interference, while reconsidering 'preconceived distinctions of signal and noise'.[6] As the performances under consideration here attest, theatre noises can become effective and affective as physical phenomena. As Rost (2011: 44) points out, the term 'noise' comes from the Latin word 'nausea', which, according to the *OED*, expresses 'a feeling of sickness' as well as 'disgust, loathing, or aversion'.[7] The somatic impact of certain noises can be related to the degree and scale of their intrusiveness. While 'noises are not intrusive per se', Rost (2011: 46-7) notes that their intrusive power consists not only in 'violating spatial boundaries, but also as transcending bodily limits in a possibly unexpected or even illegitimate manner'; in other words, the violation refers to 'the intrusion of sounds into the bodily sphere of the exposed human being'.

The violently intrusive ability of sounds is most notably observable with regard to the spectrum of sounds produced and associated with armed conflict, which J. Martin Daughtry (2015: 3) terms in his book, *Listening to War*, 'the belliphonic', which, 'through its intensity, proximity, tactility or traumatic circularity ... becomes a manifestation of violence in its own right' (2015: 271). As Daughtry shows, auditors of wartime Iraq distinguished between perceiving the belliphonic as a noise, from which they learned to tune out as much as possible to remain sane, and as a signal that provided vital tactical information about the proximity of violence. The experience of belliphonic sounds can lead

---

[6] See Jacques Attali's definition of noise as 'the term for a signal that interferes with the reception of a message by a receiver, even if the interfering signal itself has a meaning for the receiver' (Attali 1985: 27).

[7] 'nausea, n.'. OED Online. September 2021. Oxford University Press. Available online: https://www-oed-com.queens.ezp1.qub.ac.uk/view/Entry/125412 (accessed 12 September 2021).

to lingering sonic ghosts, what Augoyard and Torgue (2005: 85) describe as *anamnesis*: 'the physical recollection – literally the re-*membering* – of sound through the body' (Brown 2010: 215). This might be triggered by an actual sound or the memory of sound experienced in the aural body. This can help explain why certain sounds are perceived as intrusive by certain people while not by others: they have the capacity to trigger traumatic (sound-)memories. As Daughtry emphasizes (2015: 271), 'The pain caused by the belliphonic can be insidious, debilitating, and lasting.'

Yet Daughtry (2015: 276), importantly, also realizes that intrusive sounds can be used to create countermemories to the conflict and concludes his book with the story of Tareq and his mother 'intentionally "mishearing" the Kalashnikov fire, imagining that this sound … was actually emanating from a different source, [a Bedouin drum] the zanbur'. This ability to pacify the belliphonic violence by transforming the sounds into 'the older, richer, more nourishing cultural matrices of Iraqi musical tradition' (277) attests not only to the auditors' remarkable resilience (as discussed in Chapter 3) but also to their imaginative agency in a restrictive violent environment. This chapter seeks to develop this observation of auditors' transformative powers by suggesting that intrusive theatre noises have the capacity to alert onstage characters and offstage audiences to possibilities of transforming potential noise wounds to countermemories that can create sounds of reconciliation. As Ross Brown (2010: 218) suggests, theatre noises can have a transformative potential in the way in which they 'create doubt in the audience's mind about what is intended; to recall other corporeal feelings and make them resonate in the consciousness; to give aural glimpses of another narrative which somehow seems pertinent or indeed a parallel world'.

From dissonant harmony to sounds of reconciliations

If noise is usually associated with the negative (Hegarty 2007: 5) and, by extension, conflict, music – by contrast – is widely related to the positive and considered to bring unity and peace, and thus, as Cynthia Cohen (2007: 26) writes, is seen as 'well suited to the work of reconciling adversaries because it can facilitate communication, understanding, and empathy across differences of all kinds' (see also Urbain 2008b).[8] There is then an obvious link between the

---

[8] For a detailed overview and excellent discussion of the role of music in conflict transformation, see Phillips-Hutton 2021.

musical and the sociocultural associations of both terms (i.e. noise and music), with the concept of harmony playing a crucial role 'as a combined metaphor cum tool in processes of conflict transformation' (Korum 2020: 53). It is a widely held romantic belief that the playing of music together 'in harmony' among different ethnonationalist groups, often in the form of orchestras and other large ensembles, enables both promotion and creation of conditions for harmonious social structures. This is, for instance, the conviction that underpins the ethos of the West-Eastern Divan Orchestra, which consists of young musicians with Arab and Israeli backgrounds. As chief conductor Daniel Barenboim (2006: 3) suggests, 'Music is the common framework, their abstract language of harmony.' Yet as Gillian Howell (2018: 47) points out in her critical reflections on 'Harmony' in a special issue of *Music & Arts in Action* on 'Keywords for Music in Peacebuilding', there are notably different conceptualizations of that concept in both the musical and social domains that have different implications and outcomes for conflict transformation. For instance, Howell (2018: 47) unpacks how the notion of 'harmony-as-consonance' is musically predicated on a sonic absence of dissonant elements and directed towards resolution. While this is from a musicological perspective an incomplete concept, given that 'harmony in a technical sense includes both consonance and dissonance', this is also the case when extrapolated to peacebuilding: the muted dissonant elements can, on the one hand, be read as conflict avoidance and, on the other hand, imply the silencing of those specifically affected by the conflict who are asked to strive towards resolution and harmony, and, as a result, 'to compromise or "forgive" perpetrators without first receiving justice for their suffering. Harmony-as-consonance therefore does not necessarily correspond with delivery of justice' (52).[9]

Solveig Korum (2020: 57) develops Howell's heuristic framework to put forward the notion of 'dissonant harmony', which seeks to recognize and address the deep traumas stemming from the experience of conflict and redress the ethical shortcomings of the other conceptualizations of harmony: 'A dissonant understanding of harmony carries the prospective to name and integrate individual and sociocultural tensions and use them to improve relations between people from previously belligerent groups.' Musically, dissonance is produced by timbral alternations of 'two frequencies that are not harmoniously related'

---

[9] On the issue of compromise versus justice, see also Brewer et al. 2018 and Bloomfield 2006.

and thereby create 'a jarring sound' (58). The experience of dissonance for the listener can stem not only from dissonant chords but also from 'notes foreign to the chord' (58). The effect is resistance, irritation or repulsion, on the one hand, and curiosity and attraction, on the other. Following Rost (2011: 47, 53), intrusive theatre noises have a comparable paradoxical effect on the spectator. For Korum (2020: 58), the experience of encountering dissonant and unfamiliar sounds can 'open us up to new perspectives and creative solutions we did not previously imagine'. This chapter argues that the experience of intrusive theatre sounds, as in the above-mentioned Kabosh plays, is part of the transformative aesthetics of performances. As Fischer-Lichte (2008: 176) has theorized it, that can also generate new perspectives and behavioural patterns – and thereby give rise to and make us alert to the sounds of reconciliation that we practice and enact in everyday life in small or larger scales.

Transformative power of performed sound: Towards a politics of reconciliation

For Fischer-Lichte (2008: 157), the aesthetic experience of post-dramatic performances is 'shaped more by the experience of the liminality, instability, and elusiveness that pervades the event than by the attempts of understanding'. Drawing on the works of Victor Turner and Arnold van Gennep, Fischer-Lichte (2008: 148–9) suggests that during the performance, the perceiving subject is suspended in a state 'betwixt and between' two orders of perception: the perceptual order of presence and the perceptual order of representation. In the first order of perception, the phenomenon is perceived in its phenomenological being; in other words, as what it appears, which can trigger an unpredictable and unintentional chain of associations, of ideas, of memories, of sensations and of emotions, pertaining to reality (142–9). In the second order of perception, 'the phenomenon is perceived as a signifier that can be linked to a diverse range of signifieds' within a fictive or symbolic sphere (144).

To extrapolate this to the intrusive noises in both Kabosh performances, in the perceptual order of representation, they are perceived not so much as noises but as signs or signals that take on meaning within the context of the plays: in *Those You Pass*, the doorbell signifies that there is someone at the door; in *Green and Blue*, the explosion and the rifle shot signify the proximity of violence and thus physical danger to the onstage characters. However, the experience of being exposed to these intrusive noises interrupts this order of perception and its concomitant 'process of constituting a fictive world' by triggering in the spectator 'a stream of associations which may lead to ... auto-biographical reflection' through associated memories and emotions (Fischer-Lichte 2008: 157). The noise of the doorbell in *Those You Pass* might bring up memories of wanted and unwanted visitors alongside emotions of surprise, excitement, anxiety or dread. In turn, the sounds of the explosion and gunfire in *Green and Blue* might arouse feelings of danger, threat, angst and harm. The bomb effect was specifically created by sound designer Conan McIvor: as Kabosh's artistic director, Paula McFetridge, describes it, 'It begins with the air being sucked out (almost like a vacuum cleaner), this also serves to heighten sense of anticipation; then it is the boom and then it is a high-pitched, sustained, piercing noise that sounds almost electrical. It is followed by crows to heighten the silence before it.'[10] McFetridge explains that extreme care was taken with the sound creations of the landmine explosion and the gunshot to ensure that they were authentic but not sensationalized and also knowing 'that audiences may be retraumatized by the material'. As a result, they had a 'package of care' in delivery to manage fallout from the reaction.[11]

Within the framework of the performance, the spectator's perception does not remain within the order of presence but shifts back again to the order of representation, generating in that process new meanings and understandings of the noise, the performance and reality. As Fischer-Lichte (2008: 157) explains, 'These shifts leave the perceiving subject in a state of instability. The aesthetic experience is here largely characterized by the experience of destabilization, which suspends the perceiving subjects betwixt and between two perceptual orders. A permanent stabilization lies beyond their control.' This experience of liminality generates meaning in the form of emotions which play an active role in the creation of the performance (through the autopoiesis of the feedback loop; see Fischer-Lichte, Chapter 3). Through their internal as well as external responses (which include laughing, cheering, sobbing, groaning, crying, yawning, sleeping etc.), the spectator thus becomes actively involved in the performance as co-creator and meaning maker.

Furthermore, Fischer-Lichte (2008: 158) suggests that 'the mere strength of emotions can stimulate the impulse to act'. If these two Kabosh productions

---

[10] Personal communication with Paula McFetridge (September to October 2021).
[11] As McFetridge explain, a 'package of care' consists of all the engagement put in place to manage expectations as best as possible and to offset re-traumatizing audiences and participants.

under consideration here did not initiate more obvious audience interactions, this chapter suggests that these two productions, as well as the two others that will be discussed later in this chapter, did indeed work to generate new perspectives and meanings, which, in turn, can encourage the audience to ethical acts. More specifically, the experience of the intrusive noise associated with the doorbell and the belliphonic sounds of the explosion and the gun leads in both plays to a crisis of both onstage characters and offstage spectators that has the ability to initiate new behavioural norms and patterns, enabling multiple possible dissonant sounds of reconciliation to be heard, as will be illustrated in the following paragraphs.

Those You Pass on the Street (Kabosh, 2014): An aural space for dissonant voices

As explained at the start of this chapter, in *Those You Pass*, the initial noise of the doorbell is proceeded by a rather uncomfortable discussion and questioning of Frank by Ann, which almost leads to a conflict between Elizabeth and her sister-in-law. Yet the same doorbell rings two more times in this short play, each time alerting the onstage character(s) and offstage audience to another outside intrusion into the private sphere of Elizabeth's house. If this may generate a feeling of unease and discomfort among audience members, triggered by associations of similar past encounters and occurrences, the second ring initiates a different experience: Frank returns and, being alone this time, Elizabeth invites him inside and asks him very straightforwardly not to mention to anyone that he is helping her in his position as a Sinn Féin representative. Frank makes clear that he hears her concerns and gives her his word. If this honesty works towards building initial trust, when Frank seeks to emphasize with Elizabeth's situation by suggesting that he knows how she feels, she replies very sharply:

> I don't mean to be rude but I don't think you would know how it feels. To live amongst people and wonder if one of those you pass on the street, maybe who speaks or smiles or waves to you was the person who set your husband up, gave the information, pulled the trigger. (McKeown 2014: 15)

Nonetheless, after having clearly established her perceived difference from Frank, Elizabeth is then able to talk openly to him about their different yet also shared likes and dislikes, which include having a quiet cup of tea or coffee while listening to Classical music on BBC's Radio 3 or RTE's Lyric FM, respectively.

What neither the audience nor Elizabeth yet know is that Frank has also experienced a painful loss, when his brother was killed as an informer by the IRA and, like Elizabeth, must also 'wonder if one of those you pass on the street … pulled the trigger'. We only learn this sometime later, after the third intrusive noise of the doorbell announces the visit of Pat (played by Vincent Higgins), the local Sinn Féin representative, who was at the onset of the play openly critical, if not hostile, towards Elizabeth and her RUC family. Pat has come to inform Elizabeth about Frank's sudden death. Despite her own initial apprehension towards him, Elizabeth invites him in, and they talk about Frank and Elizabeth's dead husband.

It is important that the play gives Frank his own voice here to tell the audience directly how his brother's death affected him. But Frank is not the only one who is able to do so: throughout the performance, all four characters attain the aural theatrical space to address the audience directly to tell their stories and express their own thoughts and ideas. By making us experience the giving and hearing of such dissonant voices, we respond with our own related memories, ideas, and thoughts, before resuming to pay attention again to the voices and words on stage. In this, we actively take part in creating a webbed dialogue with present, past and fictive others. For political theorist Andrew Schaap (2005: 4), this 'willingness to engage others in a passionate and often agonistic discourse about the world we share in common' becomes the basis for an open-ended process of 'political reconciliation'.

Schaap's concept draws on Hannah Arendt's work to rework Schmitt's ideas about the ever-present possibility of the friend-enemy relation. Schaap's (2005: 4) political reconciliation redresses the shortcoming of the restorative conception of reconciliation in presupposing 'a plurality of potentially incommensurable perspectives, not only between the communities to which perpetrators and victims belong but among them'. Instead of seeking to restore vague notions of unity and harmony that avoid or transcend conflict, it embraces the jarring sounds of dissonances and takes this clash of perspectives (and perceptions) as the basis of a discourse about the world we share in common. What we experience during the performance is a commonness that is 'not yet' but that we, as perceiving, feeling spectators, are actively engaged in co-creating.[12] Consciously or unconsciously, we engage with the different and differing

---

[12] See also Dolan 2005.

worldviews and perspectives of others, on and off-stage. In addressing differences in viewpoints and questioning preconceived notions of the categories of victims and perpetrators in the context of the Northern Ireland conflict,[13] *Those You Pass* complicates beliefs in a common identity but instead posits the idea of a community as an aspiration based on open discussion and debate. For instance, in a discussion with his superior, Pat, about how best to talk to victim's families, Frank points out the importance of treating one's former antagonists not as personal enemies but as political adversaries:

> But this woman at the conference sounded sincere, and her line was, that if a republican was to meet the parents and tell them their son was killed because he was in the UDR [Ulster Defence Regiment[14]], that the UDR was a regiment in the British Army and the IRA was at war with the British Army that they'd probably accept that. But they think it was personal. (McKeown 2014: 18)

While, at the end of the play, Elizabeth questions her enjoyment of having 'a normal conversation with friends of those who killed her husband' (McKeown 2014: 33), she still insists on Pat leaving her house through the front door, in front of her husband's relatives who will most likely be outraged by seeing her friendly interaction with her former enemy: 'You're a guest in my home. You should walk out the door with me' (34).

A play that opens with the desire to attain space and recognition for one's voice and story, as expressed in Elizabeth's words, 'I'd like to speak to someone', then ends with allowing us to experience the possibility of dissonant sounds of reconciliation that can come out of such an aural space, literalizing the Latin meaning for auditorium 'as an instrument, or agency, for hearing' (Brown 2020: 47), and, here, importantly, for speaking as well. If the audience is expected to remain silent during the performance, although engaged in internal dialogues with the associations, thoughts and memories, triggered by perceptual shifts, they attain their own space to give voice to them in the feedback forms and succeeding post-show discussions that take place after the performances and are mainly chaired by McFetridge. The artistic director explains: 'with our more political work I feel it is important to offer audiences a curated opportunity

---

[13] For instance, along ethnonationalist or religious lines, given that Elizabeth's dead husband Michael was an English Catholic and member of the RUC; and that both Frank, who worked for Sinn Fein, and his brother, who was a member of the IRA, are also victims of the IRA.

[14] The Ulster Defence Regiment (UDR) was an infantry regiment of the British Army from 1970 to 1992.

to debate the themes, process the experience, air initial reactions as this often deepens the engagement. ... The discussions encourage a collectiveness, promote individuality and deepen the project impact'.

Green and Blue (Kabosh, 2016): Towards the sounds of friendship

Green and Blue starts with alternating individual fragments by the Southern Irish Garda officer, Eddie O'Halloran (played by James Doran), located on the left-hand side of the stage, and the RUC officer David McCabe (played by Vincent Higgins), located on the right hand, suggesting the 'north' side of the border. A crackling radio sound then starts to disrupt the audience directed monologues and offers a means to establish a tentative dialogue between the two men on opposing sides of the border: yet, the interfering radio noises, as well as differences in accents and sociocultural references, cause misunderstandings and miscommunications. As the two men begin to get more attuned to their dissonant voices, they start to develop camaraderie and collaborate in helping each other with regards to various border control issues, including the transporting of alcohol for a party from the North to the South.

Suddenly, however, 'the sound of a large explosion' interrupts the scene, penetrating the bodily and emotional sphere of the spectator, provoking fear and alertness and announcing danger. In the aftermath of that powerful noise, we hear Eddie's crackling voice, trying to contact his counterpart through the radio: 'EDDIE. Constable McCabe, can you read me? Over? Constable McCabe, can you read me? Over? Constable McCabe, can you read me? Over? Constable McCabe?' (McKeown 2016: 17). Importantly, the performance also creates a counter memory to the belliphonic sound that killed three of David's colleagues, with another one losing a hand and another two legs. While acknowledging this loss, and the pain and trauma that stem from it, the performance focuses on the way in which this event triggered new perspectives and new behaviour patterns between the two men. When they first speak with each other again after the explosion, Garda O'Halloran expresses his sincere concern about the RUC constable on the other side and proposes a meeting.

Their encounter comically enacts literal and metaphorical border crossings, which suspends them in the liminal space between the Northern and Southern border, which is itself undeterminable:[15]

---

[15] For a detailed discussion of liminality in the play, see also Urban 2020.

DAVID:	Where exactly is it?
EDDIE:	Where exactly is what?
DAVID:	The border?
EDDIE:	It would be a wise man could tell you that – or a farmer. Doesn't look that different over there.
DAVID:	The grass is no greener that's for sure.
EDDIE:	Well it is the same field after all.
	Eddie steps his foot hesitantly across the 'border'. He then slowly shifts the weight of his body onto the foot that is on the other side of the 'border'. He stops and looks at David. He then continues the step and finally stands with both feet in the 'north'. They look at one another. David then looks down at the 'border', then looks up. He hands his rifle to Eddie, unbuckles his belt with holster and gun and hangs it round Eddie's neck.
DAVID:	Just in case it could be regarded as an armed invasion.
	David steps into the 'south'. … They step back and forward. (McKeown 2016: 20–1)

This liminal space of the borderland temporarily releases them from their defined roles as 'Guard' or 'Peeler', offering them fresh perspectives to the ones that their uniforms confine them to. This affords them the space to share their stories about how they took on those identities, their worries, their anxieties, and their hopes. Their open and honest exchange, which respects their irreducible differences yet also recognizes their shared experiences and feeling, cements the initial gesture of their encounter through a handshake, which stands as a historically recognized symbol for reconciliation. Their interaction thus performs the model of an '*affiliative* reconciliation' that is based on friendship (Lehner 2020).

Yet just as Eddie is about to share his dream, his anticipated future is abruptly ended mid-sentence by an unexpected rifle shot. As with the prior intrusive noise of the explosion, the violent vehemence of this sound noticeably shook members of the audience who had learned to associate this noise with sensations of violation, destruction, harm and danger. As Rost (2011: 53) explains, such noises 'possess the power to provoke vigilance and alertness in the spectators, their auditory attention being intensely activated and focused on the sounds'. This heightened alertness not only to the sounds and events within the performance but also to the associations that they trigger for each spectator remains in the aftermath of the rifle shot and makes us specifically attentive to what follows. The soundscape of the play notably changes here: after the loud crack of the shot, we hear the 'kraa' sound of a raven, who traditionally symbolizes death, his shadowy image projected in flight on the large screen at the back of the stage. We then hear David's desperate voice calling his dead friend by name, knowing he is dead. A light change then transforms the scene into Constable McCabe '"*speaking*" *to a legal advisor to the RUC. We do not see or hear the legal advisor*' (McKeown 2016: 26). We hear David's official report of the incident, which, initially spoken in his own voice, is then overlaid with a looped prerecorded voice that coldly resonates through the room. When the automated voice vanishes, David, who has taken off his RUC uniform, turns to us to tell us about the changes that happened in the aftermath of Eddie's murder: with the 1998 Good Friday Agreement and the changes to policing under the Patten Commission, he was posted to Dublin to work on IT with Garda colleagues. Just as Eddie stays on as a ghostly presence on stage, David recounts an experience of *anamnesis* of his friend's words:

> One day walking through Phoenix Park, there were joggers, people on bikes, families going to the zoo, tourists; and Eddie's words came back to me. 'We're a uniform, not real people. And rightly or wrongly we now view the world from that perspective.' I stopped dead in my tracks. And in that moment I felt, for the first time that I was no longer a uniform. I could view the world from whatever perspective I decided. It was up to me. Me. David McCabe. David McCabe, person. Son of Richard and Joan McCabe. Husband to Diana. Father to Rebecca and William. (McKeown 2016: 27)

After this realization, he decided to walk the Camino trail to Santiago in Spain as a means of self-reflection, and he visits Eddie's family shortly afterwards. As Eddie is no longer able to tell his own story, David recounts to his former wife and family 'that Eddie did not die alone. That it wasn't as described in the media and official reports. I told them how he laughed and chatted and joked ... right up until the moment his life was taken' (McKeown 2016: 27). As he tells them Eddie's final words, 'One day I'm going to', David faces the ghost of Eddie on stage, suggesting a promise that his yet unvoiced aspirations and dreams will live on in him.

Transformative outside/inside perceptions

In the Image-Theatre workshop we held in Rathcoole in June 2018, participants who had seen *Green and Blue* were specifically interested in exploring the different perceptions that each of the protagonists had of each other and their status and role within the communities, where they were stationed. They recalled that neither of the men were locals: whereas Garda Eddie O'Halloran was originally from Cork, Constable David McCabe came from Newcastle, Co. Down. We decided first to create an image that captured the perception of the RUC Constable of his counterpart: the group chose to perform a sense of a harmonious, jovial close-knit community through a pub scene, where Garda O'Halloran was warmly greeted and formed an integral part. When asked to add sound, they added cheers and laughter to complement the image. The RUC man remained a visible outsider, physically and emotionally distanced from the scene, silently standing apart by himself with his gun (see Figure 5.1).

We contrasted this perception, then, with how the Guard actually experienced his own situation, which was, for him, also marked by feelings of isolation and exclusion from the local community. As Eddie tells us in the play, 'You could live here for 50 years, a hundred, and you'd still be a blow-in. Unless you've at least

Figure 5.1 The Perception of the RUC constable. Image-Theatre Workshop with members of the Rathcoole Community on 18 June 2018

Source: Photo by Stefanie Lehner.

Figure 5.2 The Garda's perspective. Image-Theatre Workshop with members of the Rathcoole Community on 18 June 2018

Source: Photo by Stefanie Lehner.

5 generations buried in the graveyard you're a blow-in' (McKeown 2016: 13). The participants reimagined the bar scene to represent the Garda's perspective of his circumstances in an image that captured him sitting noticeably sad and apart from the others, who avoid eye contact or even point at him in mockery (see Figure 5.2).

These two planes of perceptions – the imagined outside view of the Garda's life and the inside reality of it – resonate with the experience of the workshop participants who as former outside spectators of the performance now become actors and themselves perform the associations, ideas and memories that they had in their own perceptual experience of the performance, encountered during the play and in its aftermath. This experience disrupts and redraws the spectator-actor distinction: by occupying both places/roles at the same time while also fluctuating between the different orders of perception – that is, both as insider/outsider as well as through presence/representation. As Fischer-Lichte defines them, they experience a sense of liminality that opens them to new perspectives and understandings. The experience of the workshop worked to make the participants more aware of the similar experience of being marked and treated

Figure 5.3 Actor Rhodri Lewis as Ubu delivering his victory speech, inviting audience participation, in *Ubu the King* (Tinderbox, 2019)
Source: Photo by Ciaran Bagnall.

as outsiders between the putative enemies. They discussed and related to this with empathy and care, comparing this to their own experiences.

Part 2: Tinderbox's *Ubu the King* (2019) and TheatreofplucK's *So I Can Breathe This Air* (2018)

The remainder of this chapter will turn to two recent theatre productions by Tinderbox and TheatreofplucK that explore how sound can empower those confined to outsider positions who are marked as voiceless subaltern others. The Adaption Notes for Tinderbox's *Ubu the King* emphasize the eponymous protagonist's subaltern credentials: 'Ubu is of low social status and frequently tormented by the other chefs demonstrating their hierarchical status' (O'Reilly 2019: 1). Magnificently played by the Welsh actor Rhodri Lewis, Ubu is portrayed 'as a kitchen porter in a renowned patisserie', where he has to suffer 'regular hurtful remarks and beatings from the other chefs in the workplace' (1). He escapes this reality through his imagination, 'conjuring gross methods of revenge' (1). In turn, *So I Can Breathe This Air* focuses on journeys of gay ethnic minorities coming to and living in Northern Ireland: their stories detail how they are perceived and marked as 'other' – visibly and audibly most often through their minority ethnicity but also for identifying as LGBTQ+. In addition, they did not all necessarily choose to come to Northern Ireland, as some were forced to flee and arrived through the troubling systems of human migration and processes of seeking asylum.

Although the two performances are markedly different in style and subject manner, they both use the power of sound to perform states of liminality for both onstage protagonists and offstage spectators. While TheatreofplucK's production confronts us with the stories of multiple ongoing journeys, Tinderbox's production of *Ubu* exposes us to the constant oscillating shifts between Ubu's reality and grotesque fantasy. Alongside destabilizing the traditionally drawn distinctions between actors and spectators, onstage and offstage, reality and fiction and thereby also between art and life, both productions use specific sound effects and applications to capture and perform this experience of liminality: *So I Can Breathe this Air* employs high-pitched radio tuning to allow us to experience the feeling of tuning into someone else's story, memory and journey. *Ubu the King* uses distorted as well as high-pitched fading between two different soundtracks to signal the shifts between reality and fantasy. In very different yet comparable ways, sound in these productions has a powerful transformative effect which is experienced as distressing and disturbing as well as empowering and liberating. If *Ubu the King* immerses us into the soundings of conflict, *So I Can Breathe This Air* performs a space for reconciliation.

Ubu the King (Tinderbox, 2019): Sound(ing) war(s)

For a company usually dedicated to 'new writing', Tinderbox's decision to adapt Alfred Jarry's notorious play is interesting – specifically in view of the long tradition of adaptions and translations of Greek tragedies in Northern Ireland, with several of them specifically engaging with the Troubles.[16] As Mark Phelan (2016: 375) notes with reference to Shaun Richards's work (1995: 192), 'Given that tragedy is predicated on irreconcilable conflict', the attraction to use this form to reflect on the Northern Irish Troubles is 'profoundly problematic' – for

---

[16] See, for instance, Tom Paulin's *The Riot Act* (1984) and Seamus Heaney's *The Cure at Troy* (1990).

it 'removes the political agency of audiences by reproducing an enveloping, enervating fatalism which precludes meaningful political engagement or action, both on- and off-stage'. This chapter suggest that Tinderbox's decision to adapt Jarry's controversial play, by contrast, politicizes and emancipates the audiences.

To make that case, it will be helpful to provide a bit of background on Jarry's *Ubu Roi*, which famously opened and closed on tenth December 1896 in Paris, causing a riot.[17] In the words of Jane Taylor (2010: iii), *Ubu Roi* follows 'the political, military and criminal exploits of the grandiose and rapacious Ubu ... [who] attempts to seize all power for himself. The central character is notorious for his infantile engagement with his world. Ubu inhabits a domain of greedy self-gratification'. The character of Ubu has been widely adapted and inspired many different versions, especially to comment on abuses of power and human rights violations in the aftermaths of violent conflicts.[18]

However, as Richards (2008) has pointed out, there are also interesting intersections with Irish theatre history during the time of the Revival, substantiating the notion of an 'Irish Ubu'. W. B. Yeats attended the première of *Ubu Roi* and, although he reportedly did not fully understand the language, he did understand the confrontational nature of the performance and recognized the event afterwards as a theatrical revolution, which simultaneously intrigued and disturbed him: 'After ... our own verse, after the faint mixed tints of Conder, what more is possible?', he famously asked – 'After us the Savage God' (Yeats 1955: 349). Foreshadowing, in a way, what was historically politically still to come, Yeats's complex affective response can be also seen as indicative of the paradoxical parallel audience responses that Ubu productions often continue to elicit from their audience – which is also specifically the case for Tinderbox's, as will be discussed in a moment. Just 11 days after seeing the play, Yeats had his famed meeting with J. M. Synge, who was to become a renowned Irish playwright. Richards (2008: 22) emphasizes the intriguing parallels and intersections between Jarry's and Synge's works, especially Synge's *The Playboy of the Western World* (premièred in Dublin in 1907): 'What the original productions of both

---

[17] It was first conceived at the age of fifteen, in 1888, as a schoolboy satire on Jarry's old physics teacher. Jarry reworked the material into *Ubu Roi* when he was twenty-three as an unruly composite parody of Shakespeare's *Macbeth* and *Hamlet*. The play is full of scatological jokes, King Ubu's sceptre famously a toilet brush and its famous opening line ('Merde!': 'Shit!') was no doubt partly responsible for the vigour of the French public's response to its first performance.

[18] The most famous adaption of Ubu is probably Jane Taylor's, William Kentridge's and the Handspring Puppet Company's production of *Ubu and the Truth Commission*, which premièred in South Africa in 1997.

Playboy and *Ubu* have in common was an extreme reaction – indeed outrage – from their audiences.'

Jarry (1965: 83) noted in a piece published a month after Ubu's première that '[he] intended that when the curtain went up, the scene should confront the public like [an] exaggerating mirror … It is not surprising that the public should have been aghast at the sight of its ignoble other self, which it had never before been shown completely'. This notion of the 'ignoble other-self' evokes the category of the 'stage Irish': a rubric that is underpinned by colonial discourses constructing the 'Celt' as the 'ignoble' binary other to its superior Anglo-Saxon self. As a result, the category has been marked by postcolonial anxiety, and there has been a lot of vigilance about the danger of replicating these negative and harmful stereotypes on Irish stages, specifically during the Revival – as the attacks on Synge's *Playboy* testify. Yet as postcolonial critics have pointed out, such defamations and attacks often fall into a binary trap of upholding equally limiting stereotypes than they rail against. And indeed, Jarry, as well as Synge, seem well aware of the instability of binary oppositions between self and other by confronting us with our own 'ignoble other self' to show that – to reformulate Walter Benjamin's words – every civilization is underpinned by acts of barbarism.

This destabilization of dichotomies also characterizes Tinderbox's 2019 production of *Ubu*, which revives the radical potential of Jarry and Synge's work, with specific reference to our current political and social circumstances, specifically Brexit and far-right extremist voices strengthening power across the world. *Ubu the King* ran from 12 to 23 February 2019 in the Metropolitan Arts Centre in Belfast. From the moment we encounter the theatre space, we experience a sense of liminality where the clear distinctions between onstage and offstage, actor and spectator, are suspended. Before entering, the audience is asked to dress in protective clothing, which they quickly realize – when the performance begins – is very similar to that worn by the actors. The audience are thus already from the very beginning visibly made part of the performance of which they become participants, while at the same time being spectators. As mentioned before, *Ubu the King* is set in the kitchen of a renowned patisserie and Ciaran Bagnall's stage design creates a sterile, clinical environment, which encloses the performance space, as well as the audience, inside a translucent plastic curtain. This enclosed domestic atmosphere is juxtaposed with a map of the world visibly sketched onto the metallic floor, while kitchen equipment dangle from chains from the ceiling. The audience is assembled on raised platforms and seated on plastic food boxes that are used as props, slightly overlooking the gleaming metal work stations.

From the moment he enters the central stage arena, Ubu is suspiciously and disapprovingly eyed by the other chefs. He exudes a powerful mixture of childish charm and sardonic menace in the manner in which he starts to playfully make his recalcitrant presence felt by deliberately dropping kitchen equipment that he is supposed to clean.[19] Through eye contact and thus acknowledgement of our presence, Ubu directly involves the audience and, through laughter, we spurn him on and become in this way complicit in the annoyance of his coworkers. The noises he creates are clearly experienced as 'intrusive': that is, they not only disturb the actual work that is supposed to be going on in the space of the kitchen but also show the ability to breach bodily limits of both actors and spectators alike, in invading our personal 'safety zone' and making us physically and emotionally alert. As Rost (2011: 53) notes, 'The listener cannot really guard against being affected by the sounds, as they are "attacked" on a physical level.'

Ubu's initial intrusive noises alert us then to the role of sound in Tinderbox's production. While always working in conjunction with lighting, sound attains a transformative performative power: On the one hand, the change in soundscape performs the transformation of Ubu's reality in the kitchen to his grotesque fantasy world, combining in a way the social realism of *The Playboy* with the surreal absurdity of the original *Ubu Roi*. On the other hand, individually produced sounds are transformed through 'mishearing' to substantiate his fantasy of killing the king; thereby, his coworkers are transformed into puppets and accomplices, feeding him the lines to empower him to act out his fantasy. The setting of *Ubu* in a patisserie self-reflectively foregrounds processes of manipulation of materials for means of production and consumption and thereby also draws attention to the way in which the different performative elements, and sound in particular, enable Ubu first to fabricate and then to consummate his fantasy of taking over power and killing the head chef, aka the King.[20]

---

[19] See *Ubu the King* Clip 1: Opening of the Tinderbox performance from nineteenth February 2019 (2 minutes), available online: https://www.qub.ac.uk/research-centres/SoundingConflict/ResearchFindings/ReportsandPublications/ProjectMonographInformation.html (accessed 8 September 2022).

[20] See *Ubu the King* Clip 2, available online: https://www.qub.ac.uk/research-centres/SoundingConflict/ResearchFindings/ReportsandPublications/ProjectMonographInformation.html (accessed 8 September 2022).

Notably it is Ubu's staring gaze that initially brings about the shifts between reality and his fantasy, foregrounding and emphasizing that this is Ubu's perception: we are given access to his mind and we are experiencing his imaginary 'world' as he perceives it. These shifts in perception – between Ubu's subjective perception and a supposedly objective 'reality' – draw attention to the dynamic between the actual material presence of the actors, the kitchen equipment that creates the sounds and soundscape and the semiotic function they attain in Ubu's fantasy. In *The Transformative Power of Performance* (2008), Fischer-Lichte (2008: 147) introduces the term 'perceptual multistability' to describe 'the oscillating focus between the actor's corporeality and the character portrayed' – and she uses this example, then, to extrapolate these shifts in perception to the relationship between 'presence' and 'representation'. Fischer-Lichte (2008: 142) explains, 'While the phenomenon is initially perceived in its phenomenal being, it begins to be perceived as a signifier as soon as the focus strays away from the perceived object and into the realm of association. It thus becomes interlinked with ideas, memories, sensations, and emotions as signifieds.'[21] In a way, this is very much what we witness Ubu doing in this production: more and more, he perceives everything in his 'reality' as a signifier for his 'fantasy' to take over power, to kill the king: his signified. So he perceives his coworkers as either his accomplices or his enemies, while kitchen utensils turn into war equipment and tools of torture and denigration.

Ubu's perceptual shifts make us, as perceiving subjects, aware of our own implication in this creative process of meaning making – and much more than Ubu's, our perception remains in a state of flux and generates processes of association, which are grounded in past experiences and memories, that trigger associated ideas, sensations and emotions. This is specifically afforded through the use and creation of affective sounds: noises that directly affect us and stimulate our senses and emotions. The soundscape of Ubu's fantasy scenes produces feelings of anxiety, danger and harm, which several audience members associated with conflict, specifically the Troubles.[22] The opening soundtrack[23] for entering

---

[21] Fischer-Lichte's description resonates with Rancière's idea of the emancipated spectator, whose active perception generates new associations and meanings (17).
[22] See feedback forms
[23] See *Ubu the King* Clip 1: Opening of the Tinderbox performance from nineteenth February 2019 (2 minutes), available online: https://www.qub.ac.uk/research-centres/SoundingConflict/ResearchFindings/ReportsandPublications/ProjectMonographInformation.html (accessed 8 September 2022).

into Ubu's fantasy world already creates a foreboding, uncanny atmosphere that estranges us from the familiar domestic environment of the kitchen.[24] Through the cacophony, reverberations and effects of the manipulated kitchen utensils, the performance creates then a distinctly belliphonic soundscape.

This belliphonic effect is augmented in the war scene following Ubu's usurpation of power, which includes amid the loud threating vibrations of electronic drums mixed with bomb sounds also one of the noises most commonly associated with the Troubles: the drone of helicopter rotors, which is panned around the speakers to give the impression of movement and simulated in the performance through the sound of hand-crank flour sifters. However, the sonic associations with danger and anxiety are also always undercut through the comic dialogue and action; and they are further complicated in the performance's last war scene when Ubu, having massacred everyone and still hungry for more, engages in a 'full scale war'. This is sonically performed by the rhythmic juxtaposition of a threatening belliphonic soundtrack with the jaunty three to four metre of a waltz.[25] This rhythmic pattern between exposing us to a sonic warfare and the music of bourgeois pleasure arouses conflicting feelings of anxiety and pleasure, danger and desire, and enjoyment and guilt. The catchy rhythm of the waltz makes us want to join in, while the sonic warfare keeps us alert and at bay. So even though watching these war scenes obviously amplifies these ambivalent feelings, they are specifically aroused by listening to the specific sound patterns and that vigilance, as Rost (2011: 51) suggests, 'results from the audience being affected by their intrusive power'. While the visual impression allows for a certain degree of distance between the spectators and the observed actions, processes of auditory perception might directly get into the spectator's 'guts'.[26]

Several audience members emphasized on feedback forms that they were both intrusively affected and intrigued by the soundscape. In this way, we can see that the performance left its spectators in a state of instability and liminality, which, as Fischer-Lichte emphasizes, can be also experienced as a crisis due to the conflicting emotions brought forth in and by the acts of perception. This crisis is augmented by the performance's destabilization and collision of frames,

---

[24] The sounds designer Katie Richardson manipulated the sound of kitchen equipment, for instance, by filling bottles with water to achieve different notes and rotating a marble in a metal bowl.

[25] See *Ubu the King* Clip 3, available online: www.qub.ac.uk/research-centres/SoundingConflict/ResearchFindings/ReportsandPublications/ProjectMonograph2022.html.

[26] Personal communication with sound designer Katie Richardson (July 2019).

which is underpinned by its participatory elements. At crucial moments of Ubu's rise, we were asked to hail Ubu as the King, after he murdered the head chef; to applaud and cheer his inaugural 'I will make us proud again' speech, when he gained power (a rewritten version of Trump's inaugural speech; see Figure 5.3); and to get up from our seats to supply the ingredients for his celebratory orgy. It was interesting to observe audience reactions here: several members joined in vocally and participated, while others either just kept quiet or, visibly uncomfortable, averted their eyes and, in some cases, their whole bodies. However, no one directly interfered. It was thus possible to witness a range of conflicting feelings, which resonate with the ambiguously affective power of the soundscape: on the one hand, there is the desire to be part of the new 'community' that Ubu created, on the other, an awareness and guilt of being silently complicit or actively collaborating in the coup of a corrupt and greedy tyrant.

The experience of liminality, if not crisis, that the performance of Ubu generates revolves around the issue of our complicity: if we are from the very beginning already visibly made part and participant, the performance ends with an injured Ubu appealing to us for help with a medical aid kit in his outstretched arm. Thomas Rosendal Nielsen argues that this kind of theatre of complicity can provide a means of emancipation by making us aware of our processes of perceptions and thereby a self-conscious recognition of our own guilt (Nielsen 2015). Through its elements of audience participation, *Ubu the King* then reminds us of our 'shared responsibility' for what is being performatively created – an ethical and political responsibility that, as the world map on the floor reminds us, remains not just restricted to the theatre environment but must be carried out in the world.

TheatreofplucK's *So I Can Breathe This Air* (2018): Immersive listening

In a comparable, yet also notably different manner, TheatreofplucK's *So I Can Breathe This Air* makes us aware of our shared responsibility by involving its audience members as spectators, listeners, walkers, participants, performers and co-creators in its two-hour guided performative audio walk. It was part of the 2018 Belfast EastSide Arts Festival and ran from 5 to 7 August. Written by Shannon Yee and based on interviews with members of the Rainbow Project's Gay Ethnic Group (also known as GEG), *So I Can Breathe* is part of a growing body of verbatim post–Good Friday Agreement theatre productions that often commemorate specific events, experiences and stories related to the Troubles.[27] While verbatim usually insinuates authenticity and truth, the opening of *So I Can Breathe* emphasizes that it features both 'real words from interviews' as well as 'fictional words inspired by them', some of which are spoken by professional actors, while others are spoken by members of the Gay Ethnic Group (Yee 2018: 2–3), thereby deliberately destabilizing the dichotomy between life and art, reality and fiction. This is augmented by the inclusion of several performative interventions along our journey, which we however do not always recognize as such, sometimes mistaking them just as part of Belfast's street reality.

The audio journey begins in the transitional space of Belfast city centre's Europa Bus station – a place that captures the experience of liminality explored in most of the stories that we listen to: every voice tells of a different journey for different reasons – but they are all related to coming to and living in Northern Ireland and being LGBTQ+. Equipped with MP3 players and headphones, we are instructed to follow our guide (played by Noel Harron) who guides our journey and our gaze while acting out his own story about trying to contact an ominous 'agent' – a meeting that is constantly delayed. The audio walk included several other performative interventions, which add a visual reference or dramatize a specific aspect in the narratives we hear. For instance, while listening to the first story by a woman talking to us about arriving in Northern Ireland, we encounter across the street from the busstation a woman (played by shape-shifting Holly Hannaway) who is standing with her back turned towards us with a suitcase (see Figure 5.4). Listening to another narrative about the process of applying for asylum in Northern Ireland, we re-encounter her shortly later when we stop to gaze at Bryson House, where asylum seekers can record their claims. Thereby, the interventions provide an interesting contrast to the disembodied voices we listen to, which emphasize the liminality of their journeys. During the actual audio journey, they attain an ephemeral quality, and I, for one, initially did not notice them all and, instead, mistook other unstaged urban performances as such. Ultimately, as one reviewer describes it, 'We become hypervigilant, bringing all kinds of random people on the Belfast pavements into the story

---

[27] Perhaps the most famous of these have been the 'Theatre of Witness' productions of the American dancer and therapist, Teya Sepunick, in collaboration with The Playhouse in Derry. But the previous TheatreofplucK collaboration with Yee, the performed queer archive installation *Trouble*, is also based on verbatim testimonies.

Figure 5.4 Actor Noel Herron, playing our guide, directing our attention to a woman across the road, played by Holly Hannaway in TheatreofplucK's *So I Can Breathe This Air* (2018)

Source: Screenshot from film version by Will McConnell.

we're building up in our minds' (Alan in Belfast Blog 2018). This captures our active role in the creative process of content and meaning making that the immersive multisensory experience of *So I Can Breathe* generates: listening to these stories of multiple journeys, while walking through a familiar cityscape, actively transforms our perception and understanding of the city. In a way, we can understand this as a process of re-mind-mapping our known territory through the new affective story maps that we hear and create, which may rewrite and revise previous inscriptions and associations. Misha Myers uses the term 'percipient' (referring to a person who perceives the world through their senses) to foreground our active role in 'a kinaesthetic, synaeshtetic and sonaesthetic mode of perception and knowledge production' (Myers et al. 2016: 86, with reference to Myers 2011: 78).[28] Myers makes a specific reference to audio walks,

---

[28] As Myers notes, 'Where in its common usage a percipient refers to a person who perceives the world through their senses, elsewhere I have defined a percipient as "a particular kind of participant whose active, embodied and sensorial engagement alters and determines [an artistic] process and its outcomes. This mode of participation, which is led by percipients" worldviews, is distinguished from another mode of participation, which is more passive, pre-determined and/or pre-directed. It is proposed that the percipient … directs the process as they go along perceiving the encompassing environment from their bodily encounter within it; while doing so, they are making place' (2010: 67, Note 2; with reference to Myers 2008, 172–3).

where, as she notes, 'Participants may not be directly interacting with the performer' (Myers et al. 2016: 86).

In *So I Can Breathe*, there was, however, also more direct interaction with the live actors, which can 'alter the performance process and its outcomes', as Myers suggests (172–3). So one of the live performers (played by Richard Bailie) was, in the role of a participant, instructed to talk to and befriend others in the group. Close to Belfast Central train station (now called Lanyon Place), another live actor (played by Martin McDowell) attacked him (see Figure 5.5). In two performances of *So I Can Breathe*, at least one of the percipients came to his rescue and thereby directly interfered in the intervention. Furthermore, when entering the train station, all percipients were asked to fill out 'Home Office Immigration Details', which are printed on the back of their programmes. Before boarding the trains, we had to show these to the security to get through the gates: however, Richard (who did not have a programme) was refused entry as he had no paperwork. In response, in most of the performances, several participants staged their own intervention in getting upset, offering their programmes to him, and arguing with the security and the guide, while others just stood back. In a comparable albeit notably different way to Tinderbox's *Ubu the King*, TheatreofplucK's performative audio walk triggers a collision of

Figure 5.5 Actor Martin McDowell 'targeting' actor-participant Richard Bailie, with our guide Noel Herron coming in the way, in TheatreofplucK's *So I Can Breathe This Air* (2018)

Source: Screenshot from film version by Will McConnell.

frames which put several percipients in a crisis – as they had to decide whether to view the action through an aesthetical-theatrical frame or through an ethical-political one. Some resolved this crisis by following their emotional impulse to act – that is, they ethically intervened in the sociopolitical injustice they perceived, emphasizing their shared responsibility in the situation they found themselves in and actively transformed it by 'seeking out new standards of behaviour' traditionally not considered part of a theatrical performance, thereby constituting a role reversal (Fischer-Lichte 2008: 176).

This transformative potential is augmented in the way in which *So I Can Breathe* concludes. After the rather isolating experience of having listened to often traumatic stories about violence, harassment, loss, displacement and exile, our journey ended by listening to a more hopeful story expressing a positive sense of belonging that is specifically associated with the LGBTQ+ scene in Belfast. While listening, we were led into a conceptually decorated 'rainbow' space (an empty shop which was a former DUP consistency office), where the 'communitas' of the live actors as well as TheatreofplucK's artistic director, Dr Niall Rae, and the coordinator of GEG from the Rainbow Project, Dean Lee, already awaited us and invited us to share a drink, as well as our own stories, experiences and journeys of belonging with them. In providing 'a place where people come together, embodied and passionate, to share experiences of meaning making and imagination that can describe or capture fleeting intimations of a better world', *So I Can Breathe* constitutes what Jill Dolan (2005: 2) calls a 'utopian performative'.[29] In providing us with a sense of civic participation and emotional belonging, *So I Can Breathe* affords us an active role in constituting a utopian 'not yet' community that can affectively and interactively listen to, discuss, debate and contemplate possibilities and visions of different, better and more inclusive futures for Northern Ireland. In this way, *So I Can Breathe* enables us to experience and actively performs the dissonant sounds of a politics of reconciliation. Similar to the Kabosh's use of the post-show discussions, TheatreofplucK's interactive audio journey creates an aural space for giving and gaining voice; a space that, while the discussions continued, was gradually filled with the sound of dance music that invited everyone to transmute, transgressing the limits of language by communicating the feeling of togetherness through and with their bodies.

---

[29] Dolan's approach has found much resonance in readings of Northern Irish peace process performances (see, for instance, Fitzpatrick 2009 and Urban 2011).

Conclusion

The four productions discussed in this chapter use sound in different ways to trigger a crisis for both onstage actors and offstage audiences that can create new perspectives and experiences. Kabosh's *Those You Pass on the Street* and *Green and Blue*, as well as Tinderbox's *Ubu the King*, use intrusive noises to make us vigilant and perceptive of both onstage and offstage contexts: while the Kabosh productions make us specifically attuned to the sounds of reconciliation that can rise and resonate in the aftermath of the unexpected, and in the case of *Green and Blue* violent, noise intrusions associated with conflict, in *Ubu*, such noises make us aware of our own complicity in silently condoning or actively supporting acts of cruelty, war and violence. By contrast, in immersing us into the interactive listening of migrant's stories about home and belonging, *So I Can Breathe* allows us to experience Belfast in a new way, as a civic participatory space that can invoke a 'not-yet' community.

Taken together, then, these four productions delineate the liminal coordinates that have marked Northern Ireland's suspension between the haunting noises of the conflict of the past and the fragile and yet not fully realized dissonant sounds of a reconciled future community. However, the performances resist the disabling and stultifying implications accorded to this condition (Heidemann 2016); instead, the aesthetic experience of liminality works here to empower and emancipate us as active *percipients* by calling upon our affective ability to respond through what can be called 'relational listening' – an active process of meaning making generated by the experience of being suspended between two perceptual orders that enable us to *relate* our own experiences, stories, memories, thoughts and associations with what is being represented on stage by 'others' (Fischer-Lichte 2008). Thereby they emphasize our never-ceasing responsibility for shaping the world we have in common. And according to Schaap (2005), this active discussion and dialogue about a better future, which is neither founded on a politics of forgetting nor replicates the divisions that marked the past, can initiate the performative invocation of a 'we' – a 'not yet' community that is always fragile and contingent and always in a state of becoming – but that is attuned to and carefully listens to the dissonant sounds of reconciliation.

6

Working through creative practice: Socially engaged arts interventions

Pedro Rebelo

Research on how listening and sound-making practices relate to different conflict situations entails using creative methods to engage critically with materials from fieldwork and insights derived from the process. The creation of an audiovisual installation and film framed by each of the Sounding Conflict project strands is an opportunity to create links, make relationships and distil insights from the field. The work imbues an interpretative layer tinged by creative sonic and performative interplay. A critical and research-based approach to creating artwork can reveal conditions not always accessible through other modes of enquiry. The concepts of resistance, resilience, reconciliation and remediation manifest themselves in the creative act itself through dialogic practice, negotiation and collaboration. Framed within Socially Engaged Arts practice (Helguera 2011; Lacy 1995), the work discussed here aims to displace conflict scenarios into an art context in which sound, performance and storytelling explore notions of power, spatial politics and identity. The evocative effects of sound are discussed in the context of notions of identity, memory and belonging (Rebelo and Velloso 2018).

Differing from other media, installation art invites the public to immerse themselves in a designed space in which all aspects of the environment become part of the art work.[1] Multiple elements (video projection, sculptural objects and sound) inhabit a space that invites an audience to share an experience which displaces the object of observation from the media itself to an environment. The audience then become participants in the work, navigating through its narratives, aesthetics and experiential presence.

---

[1] For more on context and histories see, for example, 'Installation Art' by Oliveira (1994).

The installation *Sounding Conflict: A Performance in Five Acts* combines filmed performance, sound art, video projection mapping and sculptural design to question notions of representation and engagement (not only between the making of the work and its sources but also between the work and its audiences). A metaphorical and physical entity – a wall – is at the core of the work. Through cyclical performative action, the wall is persistently built, broken, destroyed and rebuilt by two characters who convey multiple relationships to power as each cycle unfolds. It asks questions about what a wall does to communities. Is it a protection or is it a border? Who builds a wall and for whom is it built?

In 2009 in Rio de Janeiro, the periphery of the Maré favela complex[2] was surrounded by a wall dividing the community from a major transport route between the north of the city and the centre. Presented to the residents as an acoustic barrier to mitigate traffic noise, the wall clearly became a border, a way of containing a community, separating it from other parts of the city and named by the favela residents as the *muro da vergonha* – wall of shame.[3] Since then, a wall of control and vigilance gradually became porous with broken openings facilitating the coming and goings of everyday life in the favela. These openings are acoustic connections where the 'inside' and the 'outside' blend in the ear but very much not in the eye or the mind. The Belfast 'Peace Wall' described as temporary by the British government in the 1970s not only became a permanent fixture for a generation but also grew in size and number of segments. It now stretches some twenty-one miles and continues to contribute to a segregated soundscape (Croce 2017). Varoutsos, in a soundscape study of the 'Peace Wall', comments on how it 'blocks any opportunity of cross-communication and produces disorienting effects' (2020: 123). Such effects include a resident on the Protestant side of the wall practicing Lambeg drumming[4] and it being heard through the wall on the Catholic side.[5] The significance of musical instrument

---

[2] The Maré favela complex in the North of Rio de Janeiro consists of several favela communities with an estimated population of 130,000.

[3] In 2010, the Maré-based carnival group 'Se benze que dá' (which translates as something like 'bless yourself and it'll be ok', relating to the risks of leaving home during shootings or police interventions in the favela), staged a protest entitled 'The Segregation of the Wall of Shame'. They called out the ethics of government action in spending 20 million Brazilian reais (around $10 million in 2010) in hiding the favela from the city's visitors instead of using the funds to spend in infrastructure.

[4] The Lambeg is a large drum primarily associated in Northern Ireland with its use by Unionists and the Orange Order in street parades throughout the city.

[5] In Northern Ireland, religious denomination continues to be associated with alignment with Unionism (Protestants) or Republicanism (Catholics), although this is by no means the rule. For more on the concept of identity in Northern Ireland, see McEvoy, 2008.

associations with either cultures or denominations in Northern Ireland is here made fluid (or indeed disorientating as each musical tradition is normally clearly bound to their own site and context). Although Belfast is marked by a series of wall segments as opposed to a unified wall like the one around Maré, we can draw parallels in the sense that walls in conflict situations tend to be erected by those from outside the communities where walls are to be established. An external imposition often positively presented – attenuation of traffic noise in Maré, temporary 'peace lines' in Belfast – walls represent a visual rather than a sonic border or demarcation. Sound articulates the porosity of these demarcations as waves travel through, above and beyond even so called sound barriers. In the same way in which traffic still permeates into Maré – a constant reminder of the 'other city',[6] – the Belfast wall with its sonic openings highlights the proximity of segregated communities.

In the installation and film discussed here, the ability for sound to create space and materialize action is explored through the creation of a sonic world which incessantly shifts between the concreteness of building and destruction, to surrounding urban soundscapes and references to Syrian, Brazilian and Irish hip hop. The global conventions of hip hop (in the sense of aesthetics, social function and production), shaped by very notable regional variations, present a sonic palette to question the role of music created under conflict situations. *Sounding Conflict: A Performance in Five Acts* juxtaposes and re-contextualizes the sonic realities across case study fieldwork through the use of field recordings in sound design, exploring aspects of sonic effect (Augoyard and Torgue 2005). Practices of sonic ethnographies (Smiley 2015) render audible local values and behaviours in various ways. No doubt voice and verbalization of narrative and belief play a big part in listening to a place, how a community uses language, accent, intonation and turn of phrase. Just as important is how a community makes sense, or, to be more precise, creates an acoustemology of place. Gershon (2018) articulates the role of local norms and values in relation to sound itself:

> Sonic ethnography is a construction that relies as much on local and less local sociocultural norms and values as it does on the sounds that are recorded at any given moment, the equipment that is used to conduct a recording, and the ways in which a person chooses to make an audio recording. (para. 13)

---

[6] Rio is famously divided into *morro* (hillside) and *asfalto* (concrete). Often in extremely close proximity to one another, the hillsides are where many of the favela communities live, while the concrete, referring to the built city, is home to the middle and upper classes.

In the context of socially engaged practices, these sonic ethnographies are coproduced in the sense that new understandings are developed through the relationship between artist and community. To go beyond the artist/ethnographer as onlooker and observer and the community as the observed, we must construct relations of trust, co-creation and in the context of a given project, co-dependence. Ensuring the development of ongoing relationships with participants in socially engaged projects remains a challenge for researchers and practitioners. The effects of 'parachute' projects which land on a given community – at times focused on outcome rather than process – are often brought forward as a major pitfall of these practices. The archetypical, socially engaged, site-specific artist/activist globetrotting the world spreading creative action can indeed be caricatured as a process void of contextual depth, actual participation or impact. Entangled in postcolonial discourse, questions of power, hierarchy and value emerge as fundamental issues only addressable locally through sharing and dialogue. Eurocentric frameworks embedding notions of aesthetic value and expertise can indeed *frame* or indeed contrive dynamics of participation. Creative processes can focus on activities and outcomes that are familiar and to some extent acceptable to the factions of art world while disregarding local logics and understanding. Questions around authorship lead us into the core of the relationships between participants, artists and institutions. Is the artwork by/with/for/about the community? While our archetypal artist presents their work in conferences and galleries, how do local participants and communities continue to engage with the work (or the experience of the work), if at all?

Another related critique points towards the ultimately un-reconcilable (albeit not necessarily undesirable) tension between the experience of the participant versus that of an observer:

> Today's participatory art is often at pains to emphasise process over a definitive image, concept or object. It tends to value what is invisible: a group dynamic, a social situation, a change of energy, a raised consciousness. As a result it is an art dependent on first-hand experience, and preferably over a long duration (days, months, or even years). (Bishop 2012: 6)

Very few observers, says Bishop,

> are in a position to take such an overview of long-term participatory projects – 'students and researchers are reliant on accounts provided by the artists, the curator, a handful of assistants, and if we are lucky maybe some of the participants'. (2012: 6)

Before we delve into the thinking behind *Sounding Conflict: A Performance in Five Acts*, it is important to trace back some lineage and delineate some concepts and insights that emerged from a previous project – Som da Maré (Rebelo and Velloso 2018).

Som da Maré: Reflections on a socially engaged sonic-arts practice

> Understanding sound is knowing when not to make sound. Silence can also speak.[7]

The Som da Maré project, led by Rebelo in 2014 with Matilde Meireles and Tullis Rennie, had, like many, a given timeline through which engagement, participation and creation were meant to develop and lead to some sort of public artwork. The project involved inhabitants from a cluster of favelas in Maré in the exploration of sonic arts, sense of place and identity. The project was in preparation for over one year leading to four months of weekly workshops and activities such as soundwalks, sound mapping, re-enactment, interviews, field recording and sharing of sonic experiences. Throughout the process, participants recalled everyday connections to sound memory, story and place. These experiences were in turn worked on thematically and narratively in the design of two art interventions: an exhibition in Museu da Maré and guided soundwalks in the city of Rio de Janeiro.[8] The work produced remained accessible to the public for a further two months. The project is discussed in Rebelo and Velloso (2018) in the context of participatory strategies and the soundscape as a mechanism to access everyday life experience. The ever-changing security dynamics in communities like Maré make the everyday particularly felt and intensely heard. Evaluations by participants and audience comments gathered in 2014 revealed an extremely positive response albeit framed by the excitement of having just completed a project (on the part of the participants) and touring an exhibition (on the part of the audiences). In 2018, a further point of contact

---

[7] Som da Maré participant
[8] Two types of interventions were designed to attract different publics: favela community in the Museum and publics from an affluent part of the city (Flamengo) and, most importantly, to promote exchange of publics through dedicated transport between the two which incorporated a performative element.

with Museu da Maré and the participants in the project led to a perhaps more reflective and insightful responses. Formal questionnaires conducted by Magowan with five participants, facilitated by an interpreter, generated insights, which will be discussed here.[9] The discussion attempts to take the Som da Maré project together with reflections by the participants as lines of enquiry, which together with other fieldwork discussed in Chapters 2, 3 and 5 is used as a creative and production blueprint for *Sounding Conflict: A Performance in Five Acts*. Of particular interest is the embodied language and reflective approach in talking about sound in the context of the specific conflict situation that is Maré. We continue with quotes from interviews conducted in 2018 interwoven with analysis and comment aiming to draw parallels and reveal emergent themes and approaches. The following highlights experiences in sound and music and how they demarcate time and space, how listening becomes political, an act of survival.

> Saturday is a very loud day here, there are the churches and everybody plays loud music. This relationship between community and sound needs addressed. Criminalisation of sound – impossibility of listening to music as it was with samba and with funk[10] today. I learnt to think through sound because of the project. Because these are things in our everyday, we don't understand nor value. This experience left a mark in your life. (Museu da Maré official)

> I learnt how to listen because a lot gets lost in our everyday. Sometimes there is shooting in Nova Holanda and not in Vila do João.[11] The issue of sound and learning how to listen. I can identify when shooting comes from the '*laje*' [paved area] or from the street. This is good and bad, it's a question of survival … We suffer in two ways. Conflict between factions and the police. In reality, the police, when they enter the favela they are not trying to make the situation better, they come to kill. They don't want to know who's on the streets, they arrive during school drop offs and pick ups. (a participant)

---

[9] We would like to thank Jefferson, Joyce, Alan, Mariluci, Claudia Silva and Antônio Carlos Vieira Pinto.
[10] Funk Carioca is a musical genre that emerged from the favelas of Rio de Janeiro in the 1980s, combining elements of samba, Miami bass and gangsta rap. Deeply embedded in favela politics, Funk Carioca and its subgenres place hip-hop practices at the centre of activism and social commentary (Palombini, 2013).
[11] Two adjacent favela communities in Maré.

Talking about life in the favela, a participant describes how fast everything flows and how much inhabitants need to attend to everyday. The role of mothers taking care of the children, school and university, work:

> The poor has no time to take care of her life. The poor doesn't live she survives … Each day is a different thing. I'm here today and will go home to sleep, tomorrow I wake up with a helicopter over my house, shaking my house ad I can't leave because there is shooting on the street. Everything is very fast. This project brought the idea of stopping and listening, a moment of reflection. (a participant)

> With this project we had more time to think about ourselves. We need time to stop and think. I listen to a sound that brings me back to my childhood; that brings so many good things. [Sound] brings people back to their core. (a participant)

> Sound calls people. If you want to get people to come to a show here in the museum and you do some dissemination but no one comes. You get some drums go out through the favela and in no time there are a lot of people following you. (a participant)

These reflections point towards the relationship between an intuitive response to sound, conveying information often leading to decision making (e.g. the sound of a helicopter in a specific environment), and the importance of reflection and 'time to stop'. This reflective time, often associated with active modes of listening, is here identified as valuable, while also thought of as a luxury.

One participant reflected on recalling and recording sounds of childhood and play and how a sensibility to sound led to a sense of nostalgia. Another highlighted how important it was to have their name in the exhibition and accompanying the process to completion. This participant recorded the piano passage at the end of the track 'Charlie Brown' by Coldplay. The recording was shared with the group, and it ended up being used in the exhibition overlaid with recordings of rain, a childhood memory by the same participant. This shared and personally annotated mode of listening to music represents a powerful mechanism for accessing personal memories as the participant commented:

> This track is from an album that has a story, two characters that the band created to tell a story. The sound of this music is like a walk through the city watching graffiti, driving. It's very sonic and you think you are in that car going through the sounds of the city. There's even a phrase – glow in the dark –which relates

> to this. After the climax at the end, it cuts off and this piano passage comes and I picked up this section that most people might disregard. (a participant)

One of the strategies for recalling, sharing and collating sound materials was through portable recorders given to the participants. Primarily intended to promote active listening and decision making – from the point of view of curation – this allowed for time and space with a technology relatively unknown to most. The act of sound recording becomes a form of mediating access to daily routines in both domestic and public space; the sharing of which became highly personal acts of capturing sonic moments – an acoustemological act. The identification of a sonic event through listening, memory recalling, narrative or a combination of these unfolds the recording process, which is, in turn, embedded with meaning as context gets shared verbally, as one participant noted:

> We learned how to listen better, we didn't really know how to listen ... I remember talking about sounds from our daily life, sounds of pressure pot, I brought the equipment home and recorded these sounds. I recorded my mother. Normal everyday sounds. The sound of beans in the steam pot. (a participant)

These verbal acts of sharing serve as sonic points of connection which articulate experience from the individual to the community. The reference to 'sounds of beans in the steam pot' establishes a sonic memory which becomes something that all in the group can relate to sonically, even though through different family and household contexts – an experience common to and symbolic of communities of a specific demographic in Brazil.

'At the time when the project was on, some of the things that changed were to do with the process of working with sound in the project e.g. walking in Maré with a sound recorder' (a participant). This participant remembers it was raining very heavily as he went up to the Morro do Timbau (a hill in Maré) and recorded the very heavy rain coming down. It changed how he related to spaces he walked through every day. He comments on how listening to a bird singing and the motor bike passing is a development of his perception and senses. Now, he experiences a much noisier place, whereas before he used to listen to music on headphones. He tries to listen to all the sound around; he sees there is a lot of sonic information, and he can listen to it all.

The gathering of reflections from the participants on broader understandings of sound and its relation to community revealed responses that point towards the role of listening in self-reflection and in our understanding of the world as discussed in Chapter 1 and as elaborated in the words of these three participants:

> For me it's about listening to people more ... human beings are very used to imposing what they believe in without considering the knowledge and education of others. So people need to listen to others. (a participant)
>
> [Sound and listening] can help because any action to discuss with the community about their questions has a transformative power, people discussing about their proper reality not only listening to what others are saying about them, but they are talking and listening to themselves so sound can help this reflection. (a participant)
>
> I grew up listening to shooting and endless war with screaming and death and that's agonising especially for children who don't ask to be born. So, sound is important in Maré in a negative sense because it is impossible to ignore bombs and shooting that are happening next to your front door. (a participant)

Again, the notion of listening as an active and political act emerges in contexts that range from the imposed sonic realities of conflict (bombs, shooting) to imposed ideologies and political action. The two first comments above point towards a dialogic listening which suggests giving space and acceptance to one's realities. The sense that favela voices are not listened to (in a conventional political context) is somewhat counteracted by the presence of the same voices in dominant cultural manifestations such as samba and Funk Carioca. There is, needless to say, a long history of re-appropriation of the musics and other cultural forms emerging from favela communities by the white middle classes in Brazil. One could argue that the role of samba in the carnival is a foremost example of favela cultural expression 'descending from the hillside' to be repackaged and presented in the 'city'.[12] In the same way that samba, or, for that matter, Funk Carioca shifts context when it leaves the favela, so does the broader sonic reality of resident communities and their experience of conflict. Intergenerational trauma and memory play a role here in shaping a community's listening to the everyday.

Applying Augoyard and Torgue's (2005) sonic effect terminology, we can defer patterns in the comments of the participants, denoting complex relationships to sound in everyday life. Anamnesis is the effect of reminiscence when a specific sound acts as an involuntary trigger to the revival of a memory, be it an event,

---

[12] In the case of Rio de Janeiro's carnival, it is a purpose-built track/stage which is far from the context where samba emerges but suitable for large-scale spectators and television crews, referred to as *sambódromo*. With racial politics high on display, samba's displacement through parts of the city exposes once again sound's porosity and resistance to containment. See, for example, Sheriff, 1999.

mood or atmosphere. Participant's references to the sound of rain on tin, ice cream vans, fruit vendor cries, old toys and games and of course music are some of the sonic materials provoking anamnesis. The memory itself can be recent or over a lifetime and is of course highly individual, connoting positive, neutral or negative emotions, such as specific resonances of shooting or helicopters. Here, it is important to note the specificity of the sonic material beyond its source. When these memories are deeply ingrained, the trigger for anamnesis becomes the sonic experience of something heard (rather than the naming of a given sound source). The specific echoes, reflections, localization cues and the sounding of resonating space conjure this evocative quality of sound. Although it is difficult to determine how sonic approximation plays a role here and to what extent the trigger needs to be identical to the 'original' or the remembered, there is nevertheless a specificity, we would argue, around the acoustics of the sonic event that determine the conditions for anamnesis to take place. Returning to a childhood home after it has been stripped of furniture with its inevitable change in acoustics only contributes to a sound of place as not remembered. On the other hand, a site-specific field recording of shooting in a given area is much more likely to trigger a memory than a generic shooting sound effect in the cinema. As mentioned above, when a participant is describing the ability to discern between different types of shooting, the embodied sonic knowledge, here developed out of survival instinct, is of impressive granularity in detail resulting from acute listening. As noted by Augoyard and Torgue, anamnesis has both the subjective and the individual characteristics discussed above but also has an archetypal dimension with certain sounds being associated with a given culture or atmosphere.

Complementary, and often interacting with anamnesis, is the effect of anticipation – well-known musically as the cadenza (associated with free variation) before the resolution of cadence (from the Latin *cadentia*, a falling). The anticipation of the final resolution is suspended by the cadenza but is nevertheless always present, always pre-heard. Unlike the involuntary nature of anamnesis, anticipation is framed by desire or fear associated with a given sonic experience. In armed conflict, the ever-pending sound of shooting is a strong cause of anticipation, triggered by fear or indeed by other sonic experiences which might stand for pre-shooting activity (movement, guns rattling, shouting and screaming). Anticipation suggests a state of alertness referred to by some participants when describing a state of listening out, deciphering or making sense in order to prepare themselves.

In Maré sometimes they have very strong shooting and sometimes weaker shooting but for example, instead of them stopping the shootings, they cancel the classes for the children and keep shooting not caring whether it will hit innocent people. Children don't go to school in Maré when shooting happens. They close the schools many days of the year. If we compare with Botafogo in south Rio, the quantity [of closures] is absurd. In Botafogo they have many more classes because they don't have this kind of shooting in the week. They don't stop the shooting they stop the class and the people 'suffer a lot of humiliation from the policemen and that's why the question of peace is problematic because we are not in a war. What is happening is a genocide of the black and poor population. The police coming into the community and shooting without any care. (a participant)

The four-year gap between the end of the Som da Maré project and these interviews came at a transformative time for Maré (with incessant waves of military occupation and drug cartel wars) and the participants themselves, aged between seventeen and eighteen years of age at the time of the project. This period of personal development accounts for the maturity and thoughtfulness of a lot of these statements. These recollections and reflections outline a deep acoustemological understanding of a reality that applies to Maré but also more broadly. The articulation of knowledge through sound suggests a type of understanding that differs from text or story. The recognition of what happens when we deeply listen, articulated in some of the citations above, points toward an embodied moment of understanding. Nobody can listen *for* anyone else. A personal sonic memory, accumulated, refined and curated over time, is not to be externalized. It lives and grows just like human beings, with deep interconnectedness with the environment, ever changing and exchanging.

The discussion on listening in Chapter 1 delineates various complexities and pitfalls as we attempt to model and cluster (rather than generalize) diverse experiential situations of the aural. Creative practice lives off the individual, the specific, the subjective and the fragmented. Research conducted through creative practice is by definition (if there is one) contingent (celebratory so) and very much by those who make it. The making of an art work at its fundamental is an ethical act saying *I/we made this for you*. Challenges and critiques of socially engaged art practices outlined at the beginning of the chapter are concerned with how these methods might comply with general notions of art or aesthetics. The preoccupation around how a certain practice might fit aspects of the art world can easily undermine the local and the immediately 'at hand' – the very nature of the insights one is trying to dialogue with in the first place. However valid these overarching issues are, they only invite answers from an agreed understanding of art's role in the world and indeed the function of major art institutions. Here, we focus on identifying individual and community knowledge and reflection that perhaps render these issues less urgent.

The notion of 'listening through' might be of use here. We can 'listen through' by deliberately framing our listening through the experience of another. Although we cannot determine someone else's listening (not that it would be desirable in any case), we can conceive of listening that is guided by the experience of another. Guided listening has a long history that can be traced back through soundscape studies and composition (Pauline Oliveros, Max Neuhaus, John Cage, Hildegard Westerkamp, the Futurists, etc.) not to mention ancient contemplative eastern traditions. Listening through is an exercise in directing attention to a given sound event, layer, environment, effect, gesture, texture and so on. In this sense, one is invited to focus attention on elements of sonic experience in real time. Other modes of listening through might include an interpretative layer and become part of a creative process. The example given above in which a Som da Maré participant shares a story around recording the end of a Coldplay track is an act of listening through. To listen to that recording is to listen through the participant's evocative relationships with the sonic material. In listening through, there is an act of making sense of and giving meaning to, which has powerful, multilayered ramifications for our relationship to sound. A social listening emerges as we engage in not so much the sharing of sound but the sharing of listening experiences leading to a kind of nonverbal Chinese whispers game. We explore instances of listening through as we move into discussing the creative process behind *Sounding Conflict: A Performance in Five Acts*.

Sounding Conflict: A Performance in Five Acts

The Sounding Conflict project 'investigates the effects of sound (including sonic arts, participatory music-making and storytelling in theatre) and their distribution through digital media activities' (see Sounding Conflict website).[13]

---

[13] Available online: https://www.qub.ac.uk/research-centres/SoundingConflict/ (accessed 12 September 2022).

At the core of the project 'comparative case studies with projects in the Middle East, Brazil and Northern Ireland serve as a basis for evaluating how sound is used to articulate experiences of violence, support narratives of resistance and promote peace building' (see Sounding Conflict website). To devise a creative practice response to materials and insights from these case studies stands as a key methodological strand for the project and its major public outlet in the form of an artwork. This aspect of the project was led by Rebelo in collaboration with Matilde Meireles and Patrick O'Reily (Director of Tinderbox Theatre Company). The work aimed to expressively and creatively interrelate research from across the Middle East, Brazil and Northern Ireland, to re-appropriate ethnographic fieldwork in an artistic context, to generate insights into the commonalities and differences of conflict and post-conflict situations. With a view to adding a further interpretative layer to the research materials, it was key to involve creative practitioners not directly engaged in fieldwork, which led to the collaboration with Tinderbox. As a starting point, the work had insights from Som da Maré discussed above and summaries and audiovisual materials from each research stream (Musicians Without Borders' Music Bridge Project, Northern Island; Musicians Without Borders, Lebanon; and theatres in Northern Ireland). In examining these materials, interrelationships of sound, performance and storytelling emerge and articulate unique conditions of power, politics of space and identity. Listening through these fieldwork materials became the backbone for a process of devising performative actions for filming. Filmed performance emerged as an element of the work which developed the idea of a wall as both a metaphor and a physical entity – a physical entity that would be present in the installation space in the form of bricks and sand, forming wall elements which construct and deconstruct the projection surface through video mapping.[14] The work was continued by designing a sonic environment from archival materials, field recordings, ethnographic materials and hip hop tracks from the various regions. In doing so, the process was undertaken not so much to present sounds as a representation of a given conflict situation but rather as a relational complex blend, articulating how sounds interact with each other in both space and time. As Erlmann evocatively questions, the echoes in a canyon suggest that something else is at play once a given sound source is released: 'Are there ways

---

[14] Video mapping or projection mapping is a technique which processes images to be projected on surfaces other than conventional screens, leading to a blended or augmented experience between the imagery and the physicality of the projection surface.

of documenting, analyzing, and interpreting sounds as they arise, fade away, and rebound like echoes in a canyon?' (2004: 17)

In thinking of sound design for the work, the 'echoes in the canyon' were manifest in various acoustic situations, reverberating and spatializing sonic elements through insights of the participants. At times, the design act becomes the shaping of a conversation between the abstraction of sound and the rawness of a lived story in conflict.

Before delving into the creative process, it seems appropriate to suspend flow and outline challenges that frame practice-based research in the context of this project. Although each case study 'feeding' into the art work generated vast amounts of rich materials, this compilation was achieved through disparate approaches to the collection of audiovisual materials. With each segment of fieldwork driven by its own research process, a diversity of approaches was inevitable. The international dimension of the project opened extreme potential but also significant constraints, particularly in relation to translation and interpretation, language and cultural norms. The intended international dissemination for the artwork led to an overtly nonverbal approach explored through performance and sound design. Notions of scale, representation and expectation emerge as challenges in the context of an artwork that attempts to distil and make sense of four years of fieldwork and research. Lastly, as with any artwork produced through and for a research process but also for the public, the multipurpose imperative needs addressing. How does one interweave the expectations and modes of working in art institutions, with audiences and practice-based research, with something that participating communities will recognize themselves in?

The collaborative relationship with the Tinderbox theatre company acted as another 'listening through', an interpretative turn on how the sharing of sonic ideas and understandings can act out a performative environment. Out of the collaboration emerges the notion of the wall as metaphor and its various ramifications. From the performative stance, the choreographic potential in resistance, resilience and reconciliation is played out. Patrick O'Reilly reflected on this process:

> Stepping into the landscape of rich and significantly important research collated through various pathways of the Sound in Conflict programme was a little daunting at first. It felt very important to make considerate choices in how to extract the potency of each body of research to ensure complete service to the

work in order to create a moving, breathing LIVE creative response. I felt it was much more important to work in the rich language of metaphor. We needed to work poetically to contrast the harsh and challenging reality of conflict to find a different creative breath. The next stage of the creative process was the creative embodiment. To embody a form, a concept, an ideology. We explored the process of embodiment through the space and the physical body to embody the themes of resistance, conflict and reconciliation. Using the poetic language of metaphor and the process of embodiment gave us the freedom to discover the effect of the resistance, conflict and reconciliation on numerous aspects for performance. The discovery of its effect on the bricks, the human bodies and the fragmentation of the projection mapping gave us the opportunity to have endless choices, a continuous process of resisting, being in conflict and resolving, each time something is destroyed, it is created into something else and thus we can explore a continuous pattern of a society in conflict, each one different from the other in terms of cultural influences but the same physical and psychological effect in relation to resistance, conflict and reconciliation. (O'Reilly, email communication, 2021)

The wall metaphor unfolds in various manners through the five acts of the performance. Out of walls the home emerges with its ever-present susceptibility for destruction and rebuilding. The brick as a modular element imbues the sound world and dramaturgy with multiple meanings. Through performative gesture, the brick stands for wall, home, city; the brick acts as protector, oppressor, divider … Metaphor is played out not as a mechanism for poetic imagination but rather as a device that acts out in the everyday, in our relationships with others, with place and with narratives.

The great theorists of metaphor in the Western philosophy tradition, Lakoff and Johnson (2003) have laid out space for metaphor beyond the literary device.

> The concepts that govern our thought are not just matters of the intellect. They also govern our everyday functioning, down to the most mundane details. Our concepts structure what we perceive, how we get around in the world, and how we relate to other people. (124).

Eastern thought tapped into the same sentiment long before in characteristically succinct form in a quote associated with the Buddha:[15] 'We are what we think. All

---

[15] The quote is not attributable to Buddha directly but represents an expansion on the opening words of the Dhammapada. Available online: https://tricycle.org/magazine/we-are-what-we-think/ (accessed 12 September 2022).

that we are arises with our thoughts. With our thoughts we make the world.' The working through metaphor quickly implicates issues around aesthetics. Sonic premises of time and space frame two key aesthetic decisions: the temporality of a lived experience has a special quality which is difficult to access in mediated form and an act of listening is forever framed by states of attention and therefore time – not in the sense that there is a start and finish, which is arguably the case, but, more interestingly, time as a vector of variable attention.

The metaphorical wall and its cyclical fate of construction and destruction led to an approach to temporality of action which itself embedded a cyclical quality. This eventually lays out the path for the structure of the work discussed later. The episodic nature of the filmed performative action also mirrors the cycles of everyday life – the daily frames of morning, afternoon, evening and night – which are at times implicated in conflict situations, in the ebb and flow of resistance, resilience and reconciliation. Implicated in the notion of cycle is duration and perhaps more acutely rhythm.

> No rhythm without repetition in time and in space, without reprises, without returns, in short without measure. But there is no identical absolute repetition, indefinitely. (Lefebvre 2004: 6)

Lefebvre's rhythmanalysis addresses the cyclical (as there is a need for repetition for the notion of rhythm to emerge) and the variation (the repetition itself renders itself void of interest after some time). Great worldwide musical traditions render this cyclical, as repetitions in time and space.[16]

In *Sounding Conflict: A Performance in Five Acts*, the cyclical manifests itself in a multilayered manner. From the metaphorical to the experiential, the choreographic rhythm, sound and musics gradually articulate an unfolding of repetition and variation. In tandem with learnings from cyclical musical practices and their durational implications, recurring elements must be in the here and now. This implicates an approach to sound, performance and filming, which is admittedly real time and in the present, much in contrast to conventional cinematic practices,[17] but not foreign to video installation (e.g. Bill Viola, Shirin Neshat) and practices deriving from performance art through

---

[16] Indian classical music, Tibetan chants, West African drumming and many other music traditions are structured around cyclical patterns often of overlaying scales from micro to macro.

[17] With notable exceptions in films like 'Run Lola Run', 'Russian Ark' and 'Time Code', and more recently Punch Drunk Theatre's twelve-hour, live-streamed version of 'The Third Day (Autumn)', among others.

Joseph Beuys, Hermann Nitsch and the Fluxus movement among others. This filming of action as it unfolds is primarily driven by the sonic element, the element in the audiovisual that has primordial temporality in the sense that it is vectorized (Chion 1994), that cannot go back nor fold on itself without alerting a shift in our reception of a work. The 'real-time' filming of a choreographed action is then a sonic performance that is suspended, only to be rooted by the fleeting audiovisual relations of synchresis, when two, however disparate sonic and visual, events occurring at the same time come together to form a 'third' event (Chion 1994). The choice of visual perspective for filming was that of an observer. The observer takes on a fixed central perspective that becomes the position of the camera and the location of a viewer of the projection in an installation setup.

Sound design was a driving force behind the creation of the work, as sonic materials from fieldwork are shared, manipulated and 'listened through' to create an immersive soundscape with emphasis on the non-diagetic and site-culture specific. Sound materials include field recordings from Lebanon (Zerieh 17 August, Zerieh Pop Up Concert); BBC Sound Archive (Belfast riots recordings); field recording environments from Rio de Janeiro, Lebanon, Abu Dhabi and Northern Ireland; recordings from the Belfast Soundmap;[18] and location (studio) sound, sound design, foley and hip hop samples: Kneecap 2019, The Synaptic 2018, Somos CV and MC Orelha 2014.

The use of hip hop reflects its ubiquitous presence across all regions in the project and indeed the world while maintaining a remarkable degree of local specificity.

> The site specific forms of hip hop and rap in Jordan respond to specific milieus that produce unanticipated dynamics in the sounds associated with the genre. This is because they are coupled to improvisatory practices that arise within a very specific political context. (Milton-Edwards 2021: 76–100)

Hip-hop practices notoriously implicate societal dynamics in both local and global manifestations, though by no means generalizable. As Rollefson warns us:

> In continuing to be dazzled by hip hop's globalizing novelty as it expresses new collisions of local and global cultures we have a tendency to buy into the narrative that this thing called 'globalization' is something new and unprecedented. As the postcolonial frame continually reminds us, it is not. (Rollefson 2017: 3)

---

[18] Available online: http://www.belfastsoundmap.org/ (accessed 12 September 2022).

Figure 6.1 Sounding Conflict: A Performance in Five Acts: Still from Act I

The multilayered, sampled and re-appropriated approach to much hip hop links in fertile ways with the notion of 'listening through', the sample as the iconic sonic gene that is forever re-contextualized.

> In parallel with the lyrics and visuals, the regular, metronomic beat of the BBC News theme tune in *Get Your Brits Out* represents a form of hegemonic authority (Van Leeuwen 1998) that of British state television – which Kneecap subvert by re-purposing it as a backing track for a song rebelling against the authority of the British state. (Ó hIr and Strange 2021)

Structurally, the performance is divided into five acts lasting just under thirty minutes in total, each framed by a quote from fieldwork as follows (see Figures 6.1 to 6.5):

Act I: Resistance (0'00"–6'59")

> People will often blame themselves, but when they hear someone else's story, they see it as a social problem that needs to be addressed, so it's not a matter of something between me and myself, it becomes a story of the community and it mobilises people to different causes.
>
> Farah Wardani 2020, Interview with Julie Norman, Beirut: Lebanon (on Playback Theatre in Lebanon. Farah is the founder of the Laban Theatre Company and an actor/director)

Figure 6.2 Sounding Conflict: A Performance in Five Acts: Still from Act II

Figure 6.3 Sounding Conflict: A Performance in Five Acts: Still from Act III

Act II: Reconciliation (6′59″–10′36″)

The sight and sound of free-flowing vehicles crossing the Border into Northern Ireland conjure a feeling of peacefulness, harmony and freedom. This is how people want to live.

Annette McNelis, Musicians Without Borders

Figure 6.4 Sounding Conflict: A Performance in Five Acts: Still from Act IV

Figure 6.5 Sounding Conflict: A Performance in Five Acts: Still from Act V

Act III: Resilience (10′36″–17′11″)

When you don't have freedom, at least music can help you express, can help you break the boundaries, break the borders, that are around us.
>>Shafiq 2017, Interview with Julie Norman, Bethlehem: Palestine,
>>23 March (on Musicians Without Borders in Palestine.
>>Shafiq is a youth trainer)

Act IV: Resilience (17'11"–20'37")

And they said, you gave us our children back.

From Sounds of Palestine staff

Act V: Resistance (20'37"–24'19")

Peace needs to be constructed at the micro level

Jefferson, Brazil, Som da Maré participant

Act VI: Coda (24'19"–27'49")

The filmed performance is intended for presentation as an installation which is adapted to a given exhibition site and includes live performance interventions. The projection of the film includes distortions, shadows and fragmentation caused by the physical bricks in the space. These bricks are used performatively in dialogue with the actions on film. Local performative responses act as a fundamental way of reinterpreting and re-contextualizing the filmed performance. For the first showcase of the installation (June 2022) in Museu da Maré, Rio de Janeiro (see Figure 6.6), performers Geandra Nobre and Matheus Frazão intervened by introducing a local reference to a wall – in this case, an invisible wall which divides the Maré favelas into territories controlled by opposing drug gangs. This layering of local references embeds the work with a dynamic and iterative nature which continues the research process from which the work emerges.

Sounding Conflict: A Performance in Five Acts is both a researched artwork and research process in itself. Acts of listening through unfold further as the work is experienced by audiences. We have outlined how findings from Som da Maré and fieldwork have framed the creation of *Sounding Conflict*. This framing is often understated rather than the overt, suggestive rather than determining. A work like this must create a space of encounter which, given the subject at hand, is full of fragmented narratives and contradictions, some to be reconciled when experiencing the work others left unresolved. The reworking and listening through sounds and stories of conflict does not produce a coherent hegemonic narrative. Why should it? It does, however provide opportunities for rooting experience in what art practice is best at: making the specific big. From local,

Figure 6.6 Sounding Conflict: A Performance in Five Acts: Live performance in Museu da Maré, June 2022

often, individual situations, the sounds of someone's everyday can open a field of exploration in which identity, memory and belonging come into proximity or retreat into the distance given each listening through. Resistance, resilience and reconciliation are sonically interplayed, heard through moments of aggression, stillness, endeavour, effort, joy and fear. Our everyday sound experience is forever linked to our responses, and hence sound is a privileged modality for accessing the visceral, unspoken bodily reactions that are only too present in situations of conflict. Sounding conflict is then not so much an act of sounding but an act of listening, to one's own world and to that of others.

Conclusion

Fiona Magowan, Julie M. Norman, Ariana Phillips-Hutton,
Stefanie Lehner and Pedro Rebelo

Our intertwined analytic framework of resistance, resilience, reconciliation and remediation calls for greater attention to the significance of sonic practices in verbal and nonverbal creativities and their somatic effects to understand how they generate collaborative, transformative outcomes for interdisciplinary conflict research and creative practice. We have shown how participatory arts are employed by musicians, arts facilitators, theatre practitioners, community activists and other stakeholders as a means of 'strategic creativity', with a range of aims, including transforming trauma, promoting healing and facilitating empowerment. Yet, in tandem with interpersonal development, our research studies also highlight how extraneous political agendas can influence participatory perspectives, raising questions about the extent to which different kinds of creativity or creative programmes can and should address the political and under what circumstances.

It is clear that in both conflict and post-conflict contexts socially engaged creative practices can open up unique opportunities for the expression of resistance, resilience and reconciliation broadly defined as well as their remediation into new forms. At the same time, there are limitations to strategic creativity that require critical engagement and reflection. In this conclusion, we discuss how each of the elements are variously interwoven throughout the case studies and elaborate upon the observed benefits and limitations of participatory sonic practices in furthering creative engagement in conflict and post-conflict settings.

Towards an ethics of listening?

By making listening our key method and methodology, this book tentatively traces what could be called an 'ethics of listening': it enables both a listening *to* and a listening *in* to other sounds, voices, narratives and stories of violence and conflict of resistance, resilience, reconciliation and remediation. This kind of listening, we suggest, has the capacity to deepen and foreground our engagement with the world we share in common: our response*ability* for transforming conflict and building peace. As Salomé Voegelin argues, listening needs to be understood as an active 'act of engaging with the world' such that 'the listener [is] intersubjectively constituted in perception, while producing the very thing [s]he perceives' (2010: 3, xii). This emphasizes the active role of each of us as co-composers in shaping the world in which we live. As Voegelin continues, 'In listening I imagine the world' (9), and it is this creative act that forms an important part of what John Paul Lederach (2005) and others term 'the moral imagination' (see also Cohen et al. 2011).

Lederach (2005: ix) defines the moral imagination as 'the capacity to imagine something rooted in the challenges of the real world yet capable of giving birth to that which does not yet exist', which is crucial in transforming violence and 'building constructive social change in settings of deep-rooted conflict'. According to Lederach, the moral imagination is generated by the following four capacities: (1) the ability to imagine ourselves in a network of relationships, including our enemies, (2) the ability to embrace complexity that refuses polarities and divisions, (3) an open commitment to the creative act, and (4) a willingness to take the risk of encountering the not yet known. These four faculties have been mobilized by the listening processes generated by our case studies. In listening *to* the performed sounds and narrated stories of perpetrators and victims, or by listening *in* to conversations and debates between enemies and friends, we develop more complex understandings of each other in appreciating each other's humanity and presence in the world.

Silence plays a crucial role in this process. It is noteworthy that the most intense processes of listening in the theatre productions discussed in Chapter 5 followed a momentary absence of sound that itself often followed the intensity of intrusive noises. In the two Kabosh plays, for example, the silence after the intruding sound of the doorbell in *Those You Pass on the Streets* and the bomb explosion in *Green and Blue* enabled a process of deep intersubjective listening. In silence, the individual's body and subjectivity become an echo chamber not only for sound production, as Voegelin (2010: 83) asserts, but also for reception to the stories and narratives of others: enemies, strangers and friends. These sounds and silences convey emotions that resonate with our own, sounding within and without, thereby emphasizing interconnectedness. This kind of ethical listening resists pre-given binary categories and divisions, such as the supposedly congealed roles of perpetrator and victim addressed in *Those You Pass*. By opening themselves to the words and sounds of the other, the characters in the play discover similarities as well as differences – as we do as audience members with resonant stories and experiences on- and offstage.

Moreover, in and through temporary silence, we become aware of our reciprocal roles as both listener and sound maker (Voegelin 2010: 84). In the theatre performances discussed in Chapter 5, this was foregrounded through audience participation, such as when the audience for Tinderbox's production of *Ubu the King* contribute to Ubu's noise making by cheering his inaugural speech with their own voice. While noise isolates 'the listener in the heard' (44), our active contribution and complicity in its production connects us with the others who are possible simultaneous noise makers and noise listeners. In this way, noise amplifies not only 'the struggle for identity and space' but also confronts us with a variety of 'social relations' (45). This holds both for the character of Ubu onstage as well as the audience, who are spatially included in the scene. In the abrupt silence that follows Ubu's injury and his appeal for help, gesturing to us with a first-aid kit in his outstretched hands, we are reminded of our responsibility – not merely in terms of practically helping him but of our ability to respond to his story by acknowledging his apparent anger and frustration. In listening, we recognize in him a more complex character than that of a power-hungry tyrant. Although the audience's willingness to take an actual risk to intervene was not taken in this production, the post-show discussion confirmed that the majority of the spectators realized that complexity of the character.

In the TheatreofplucK's audio-walk, *So I Can Breathe This Air*, the percipient's immersion in the sounds of the prerecorded voices alongside the present sounds of the city enabled us to creatively co-create our own soundscape through acoustic ecology: the various sounds, accents and stories of multiple belonging worked to broaden and transform our perception of Belfast. Here, the risk of the unknown was evident not only in the ethical intervention of some audience members to save the purported victim of a staged attack but also in entering the final confined rainbow space where we were invited to share our experiences, visions, fears and hopes of peace and an inclusive community.

In facilitating a quasi-Lederachian ethics of listening, the sonic practices we detail here affect and produce what is audible and visible, and thereby known and understood. According to Jacques Rancière (2004b), it is exactly this that makes out 'the politics of aesthetics': more specifically, 'politics is aesthetic in that it makes visible what had been excluded from a perceptual field, and in that it makes audible what used to be inaudible' (226). The sonic and auditory dispositions of these practices have shown the potential to impact, and thereby transform, our sensibilities: our ability to sense and imagine life beyond conflict, divisions, polarities and binaries.

Musicking as ethical reasoning and care

Our research with Musicians Without Borders (MWB) has demonstrated the benefits of a collaborative participatory action approach, whereby our objectives and activities were collectively analysed at different points through individual and group review processes and critical reflection (see, e.g. Odena and Cabrera 2006; Reason and Bradbury 2007). It is apparent that the most effective outcomes of creative practices, such as participatory musicking, depend upon having a protected arena in which to analyse ethical value processes in verbal and nonverbal interaction, whether as facilitators or as trainees. Since each person is a product of the environment in which these musical activities happen, there is a continual need for self-regulation and a keen awareness of one's preconceived ideas and ways of working in terms of the impacts that they can have on the desired outcomes. This process of cooperation in musical expression thus lends itself to engaging in shared goals that require sustained input to achieve transformative outcomes of attitude and behaviour.

An augmentation of trusting relations through musicking is especially important as research has demonstrated that teachers feel generally unable to deal with educational situations that may elicit tension (Hughes 2007) and neither are educational changes in the guidelines sufficient to address attitudinal changes to longstanding intergroup conflicts (Smith and Robinson 1996). However, as we have seen in this volume, participatory musicking enables a field of interaction to be developed that enhances the potential for an equivalence of ethical values and social capital to be distributed among participants. The term 'social capital' relates to 'the value of being connected, the specific benefits that flow from the trust, reciprocity, information and co-operation associated with social networks' (Dowling 2008: 184). Nonetheless, rather than retain the neoliberal connotations and echoes of 'capital' as something inert or commodified, we refer instead in our analysis to social relations of capital as something that musicking does or produces in the interstices of action. How people interact with one another, for example, in processes of song writing and the rhythmic timing of call and response singing all require a deliberative and intentional mode of connection, attention to one another's needs and a positive response. Thus, in listening to each other, relational capital accrues intersubjectively. The acquisition of skilled interconnections is necessary for the wider maintenance and implementation of other kinds of social networks, in which there is an accumulation of other kinds of capital (e.g. financial, political and material).

These interstices of action raise broader questions about the nature of connectedness in different ways; how are they manifest and 'through what effective cultural and aesthetic means' (Dowling 2008: 190) are they produced? In the process of informal musicking, there is a decentring of the facilitator and re-centring of the participant as the subject of musical production. In this nexus, each musical participant is variously drawn into an arena of 'ethical reasoning ability' according to the merits of their potential actions, (or *phronēsis*, following Aristotle) (Bartels 2019: 38). The choices that are made in this context enable each participant to negotiate their knowledge as well as the outworkings of 'good actions' (39). The displacement of decision making from the facilitator to the participant opens up possibilities for different kinds of ethical action which will further inform their decision making in other areas of their lives. Thus, rather than the music facilitator teaching what 'ought to be done' in any given musical situation, participants are invited to reflect individually and jointly upon how their reactions can collectively create the best possible outcomes. Indeed, what counts in the learning process and feeds into attitudes towards social justice gains is the opportunity for each person to be responsible for their own aesthetic, musical and interpersonal judgement as the musical piece is brought forth through creative and interpretive dilemmas that musicians hold, at the same time they must make decisions on how to act in the present to engage with the realization of the musical piece. These actions are circumscribed by three modalities of praxis: 'perception (*aisthēsis*), thinking (*nous*) and striving (*orexis*)' (43), which Bartels argues come together in musical creation and inform not only ethical reasoning but also the quality of the aesthetic production. In recognizing different forms of quality, participants also reflect upon their own contributions and thus strive to create their own sense of what is 'good'. This interlinked process brings out the dynamics of virtue ethics through 'the personal integrity – the character, if you will – of the actor or ethical agent. On this view, one's inclinations to right action stem from who one is (or, perhaps more precisely, who one is in the process of becoming)' (Bowman 2016: 68). These musical interactions then build up capacity not only within each participant but also between facilitators and participants acting together to realize their potentialities at the same time as they also further transform how they perceive themselves and others in various aspects of their daily lives.

In an arena of safety and non-judgemental musical practice, an ethics of care can be clearly manifest as it stems from these principles of ethical reasoning. The facilitators of MWB are skilled practitioners in this ethics of care which involves four aspects identified by Noddings (2002: 287) as 'modelling, dialogue, practice and confirmation'. On the one hand, facilitators are role models for the participants, while they also model qualities of attitude and conduct in performative action that participants are then able to follow. Equally, the facilitator seeks to empathize with the level of experience, concerns and insecurities that participants may be feeling during the tasks they are asked to perform. In this sense, modelling is about assisting others to bring out the optimum they are capable of while recognizing their limitations. In doing so, relationships are deepened and senses of trust are increased mutually between facilitator and participant. Dialogue is integral to this process which, in the case of MWB, can be verbal or nonverbal, though the essence of it is that it 'is a common search for understanding, empathy, or appreciation. It can be playful or serious, logical or imaginative, goal or process oriented, but it is always a genuine quest for something undetermined at the beginning' (Noddings 2005: 23). It is critical that participants understand the value of each other's contributions, supporting one another equally in the process. To enhance the dynamics of the outcomes, facilitators need to 'be able to envision imaginary spaces' in which trainees can utilize their skills (Juntunen et al 2014: 252) and to be able to generate 'particular social-musical spaces as potential *future possibilities* ... [and] nurture the idea of 'transformative music engagement' (255). The effect of dialogue then is to realize an expansive arena of creative acceptance, rather than an imposed setting, in which a wide variety of musical actions and interactions are possible. Such envisioning of spaces of acceptance, collaboration and flexibility are also a form of care in practice for another person. By recognizing the need for care, facilitators and participants alike can affirm and confirm their efforts to be creative as they connect in caring ways with one another in and beyond their musical practices.

As we have seen, the approach of MWB to resilience is ingrained in their ethos of nonviolence that embraces an elicitive framework for peacebuilding whereby 'everyone teaches and learns, so leadership is shared; learners' experiences and concerns are valued; there is a high level of interactive participation; people co-create new knowledge and engage in critical reflection [through their creative practice]; there is a connection made between the local and the global; and people work together for change' (Shank and Schirch 2008: 232). The link between learning specific musical techniques in a safe space and then being able to put them into practice in real contexts enhanced participants' senses of resilience as they found out that they had to deal quickly and efficiently with issues as they arose and use their creative toolkit to deflect and manage any difficulties arising.

As some of the creative practice of MWB is nonverbal, the participants were also learning to watch, listen to and respond to 'the symbolic channels of facial expression, body posture and eye movement ... channels that carry important information about emotions, energy and thought' (Shank and Schirch 2008: 235). Understanding how the locus of the body can be one of the greatest assets in the peacebuilding process is central to MWB's artistic practice. By developing nonverbal creative skills in embodied interaction, practitioners are able to read, to listen to and to lead one another in a multitude of expressions and emotional ranges that can transform how they understand communication, relationality and their interactions with the world around them. They thus acquire a new framework for analysing attitudes, behaviours and expectations of themselves and others that can be put to positive effect in other musical, artistic or creative contexts.

Narrating empathy

The capacity of participatory sonic actions and musicking to encourage trusting relations and dialogue between individuals and groups is closely linked to the question of the capacity of these activities to encourage feelings of empathetic connectedness between individuals or groups of people. Empathy, understood here as an imaginative perspective taking shaped by cognitive and affective processes, has been productively explored within fields such as psychology, cognition, the medical humanities and aesthetics alongside its theorization within studies of the participatory arts and music (Cao et al. 2021; King and Waddington 2017; Thompson 2001). Moreover, organizations working to transform conflict have promoted it vigorously as an 'invaluable natural resource' for such activities (Empathy for Peace 2019: 4). Yet the connection between empathy and emotion has meant that it has been taken up only slowly within other fields associated with conflict transformation, such as international relations and politics (Head 2012). This is surprising given the widespread agreement that beliefs, feelings and interpersonal relationships are crucial factors shaping sociopolitical outcomes, but it perhaps reflects the intimate concerns of these fields with explicating the nature of power – including the potential malevolent power of leverage generated by empathetic understanding. If empathy is a resource, it is one that can be used in multiple – and possibly conflicting – ways. As Ariana Phillips-Hutton (2020) has argued, this means that we should resist overly optimistic characterizations of the relationships between empathy and music or other sonic practices in favour of a grounded exploration of how these practices contribute to a deeper understanding of our interrelatedness.

Themes of empathy and understanding arose in several of our case studies, but it formed a particular point of interest throughout our discussions with Kouyoumdjian, Yousufi and members of the Buffalo String Works (Chapter 4). In these instances, increasing audience empathy for those individuals whose stories were part of the musicking and for the wider refugee community figured prominently as a desired outcome of both the storytelling itself and of the consequent compositional and performance process. As several interviewees indicated, the use of identifiable human voices (in some cases localizable to specific human beings) and the inclusion of intimate details – all accompanied by music – was designed to encourage the audience to identify with the storyteller's humanity and, subsequently, to act differently towards refugee populations in general. Likewise, the use of personal stories and 'sonic memories' (Diamond 2015: 276) within theatrical productions in Lebanon and Northern Ireland are predicated on the capacity of this artistic storytelling to generate a deeper understanding of disparate individuals and groups. This potential for interpersonal and discursive change and exchange can be facilitated by the 'sonic subjectivity' of 'stage props' (Brown 2020). In turn, this raises questions about how the staging and performance of 'kinesthetic empathy' (Reynolds and Reason 2012) can impact the moral imagination.

This widespread deployment of sonically mediated empathy draws on a range of widely accepted beliefs about the capacity of the arts to encourage empathetic identification. As Felicity Laurence (2017: 26–7) has suggested with regard to musical empathy, research in this area has frequently elided the empathizing that happens within a given artistic experience with a more general, 'social' empathic response. However, the mechanism for transferring one experience of empathy to another remains poorly understood. Moreover, as our research shows, the potential for the sonic participatory arts to develop empathy cannot be disentangled from questions of interpersonal relations or from the potential abuses of power within such narrative and ethnographic relationships. The serious ethical issues that haunt ethnographic research are equally evident in Kouyoumdjian's rejection of speaking on behalf of those whose stories envoice her ethnographic compositions. As this demonstrates, the question is not merely about who speaks and who listens, but about whose stories are told, in what way and by whom?

These are questions that echo across our case studies, from the shaping influence of large charitable organizations such as MWB on structures of resilience (Chapter 3), to the use of testimony for reconciliation in Northern Irish theatres (Chapter 5) and the relationships between the sound installation and community members in the Maré *favela* (Chapter 6). Moreover, these are questions that animate key contemporary debates over the possibility of 'decolonizing' empathy (Pedwell 2016), reframing pedagogy (Zembylas 2018) and shaping the direction of ethnographic study of the participatory arts (Enria 2016). One element emerging from the work detailed in this book is the need to move beyond formulating voice as something of which to be either fully possessed or dispossessed. Whether individuals are living in the middle of conflict, in post-conflict societies or as refugees, giving voice (and the agency often presumed to accompany it) is always an act of relational dialogue. The choice to tell certain stories and not others – the strategic use of what we might call, after Semhar Haile (2020), refugee (in)audibility – is an assertion of agency that rejects a binary of either victimhood or activism.

The stories we tell move among us to shape our understanding of each other and of the world in which we live. One of the key areas of future research that emerges from this discussion of narrative and empathy is the capacity of narratives that are told and retold to develop empathic relationships across significant gaps in space and time. In a world in which conflicts and their consequences are regularly screened for, and yet infrequently felt by those in positions of significant social, political and cultural power, understanding how the sonic arts might encourage remedial action is of significant and, growing, importance.

Power to truth versus truth to power?

Socially engaged creative practices can create unique opportunities for resistance, resilience, reconciliation and remediation in conflict and post-conflict contexts.

Sometimes these processes are direct, but in our research, we found that they were more often indirect, echoing and informing each other. In each of these cases, sonic practices facilitated the emergence of literal dissonances that sought to enhance social cohesion and challenge the political status quo. But did they? On the one hand, socially engaged creative practices like sound, music, theatre and storytelling can create powerful horizontal change by transforming how community members engage and interact with each other. But on the other hand, there are limitations in influencing vertical power structures and altering macro conflict dynamics that require critical engagement and reflection.

In the case study of Lebanon, for example, the sonic practices of storytelling and theatre contributed to individual and community practices of resistance, both direct and indirect. Ex-combatants from Fighters for Peace (FFP) and actors and activists from Laban Theatre used their stories and performances respectively to creatively contribute to the 2019–20 revolution. Through maintaining a physical presence in public spaces, they lent their voices to the broader societal calls for change and accountability. The primary value of the participatory aspects of their sonic practices, however, was in the form of indirect resistance. FFP members didn't just tell their experiences from the civil war; instead, they used their stories as a catalyst for engaging others, including younger activists, about processes of political change, the moral and strategic limitations of violence, and the philosophy of community and solidarity. Likewise, Laban didn't just perform to static audiences, but rather they used their performances as a starting point for conversations about individual and community grievances, fears, hopes and reflections. These initiatives blurred the lines between resistance and resilience and reflected the previous work of both organizations, especially in terms of their engagement with the legacy of the Lebanese civil war. In a context in which accountability was scarce and the state encouraged a kind of collective amnesia, simply talking about the civil war openly and honestly could be considered a form of indirect resistance.

FFP's work enhanced resilience and reconciliation among the members sharing their experiences, as well as of those hearing their stories and engaging in subsequent dialogue. The members themselves went through a deep self-reflection process, using oral history techniques to reckon with their past actions during the civil war and to acknowledge and voice traumas from that part of their lives, cultivating resilience for dealing with the past. They also engaged in private dialogues with each other and other ex-combatants, often from different sectarian and geographic backgrounds, allowing for rare opportunities for identifying shared experiences and reconciling some differences. Further, their openness in sharing their experiences with others in the community created spaces for others to likewise grapple with their experiences and actions during the civil war, and at times, if not reconcile, at least acknowledge the humanity of those who fought on other sides of the war. FFP's outreach to youth, especially in areas where many young people are drawn into sectarian groups utilizing violence or extremism, was also an example of using stories and dialogue to challenge notions of strategic resistance and build resilience among youth. Indeed, because many FFP members were minors themselves when they joined armed groups in the civil war, their personal stories brought a degree of salience to youth groups as they explained how their own experiences revealed that violence is not the only way to fight injustice.

Laban members also viewed their work as much broader than generating resistance, emphasizing that their performances sought to create a space for discussing difficult topics and identifying commonalities among community members. The creative nature of their theatre, even when dealing with serious subjects, allowed community members to engage with the topic in ways that were more accessible and approachable than a formal discussion. Performances also enabled actors and audience members alike to identify areas of shared trauma or vulnerability, as well as common sources of strength and agency, developing both personal and collective resilience. Further, the conversations inspired by the performances, like FFP's dialogue sessions, created openings for engaging with others, increasing understanding and reconciling with others or even with oneself or with past experiences.

In these case studies and in others, we see the social solidarity concept of the 'red thread' echoing throughout the sonic processes. As noted in Chapter 2, Laban uses the idea of a binding thread or line that is present in every performance and brings everyone together. This concept is useful outside the theatre space for envisioning the sense of empathy, trust and shared experience that can be cultivated through creative practices, especially those that engage with difficult stories or expressions of trauma, conflict and war.

Our other case studies also reflect how each of our core themes is interrelated. In Brazil for example, participatory sound interventions helped develop a sense of belonging and identity in the Maré *favela* complex, thereby countering attempts by authorities to isolate, divide and marginalize the community. While fostering community resilience, sonic practices also functioned as a form of resistance to the status quo by challenging physical barriers that separated the favela from other parts of the city, just as activists have sought to use art to counter the geographical sectarian divides in Beirut. A similar breaking down of barriers between different lived spheres through artistic remediation is evident in the stories told by refugees in Chapter 4.

In Northern Ireland, the theatre companies discussed in Chapter 5 also facilitate a process for building trust and community between actors and audience, even though reconciliation is still a contested term for many. However, as with Laban in Lebanon, the theatre provides a rare space for engaging in critical conversations about the conflict, identity and belonging. This is especially true in participatory and immersive productions, in which audience members both experience and share stories related to the conflict, facilitating processes, discussions and debates around dealing with the past and envisioning a better community. In this way, reconciliation is not so much about healing or forgiveness as it is about confronting the past as well as the present.

Likewise, the work of MWB in Northern Ireland that appears in Chapter 3 focuses less on reconciliation and more on using music to build resilient communities and individuals. Like FFP and Laban, most of MWB's work counters dominant assumptions about divided communities by using a creative practice to facilitate a shared space for community healing and solidarity. FFP especially focuses on individual trauma recovery, with sound-based interventions as a means of creatively reckoning with the past, resisting social and political pressures to bury or forget the memories of the conflict.

As the ongoing conflict in Palestine renders ideals of reconciliation not fully viable, MWB also concentrates on resilience. However, MWB bridges divides between refugees and urbanites, Christian and Muslim, and middle-income and lower-income groups within the Palestinian community, thus resisting some social norms and building more resilient communities. The programme in Palestine also focuses on healing from trauma, especially for youth, and offering young people a sense of purpose. This sense of building resilience is different from resistance in the traditional sense, but it reflects well the Palestinian concept of *sumoud*, or steadfastness. By fostering a sense of perseverance, pride and solidarity among youth, MWB contributes to this spirit of everyday resistance.

Creative strategies for the creative arts

While all the cases that we have observed reflected elements of resistance, resilience, reconciliation and remediation that were strengthened or enhanced by socially engaged sonic practices, there were limitations to these interventions as well. First was simply the scale and scope of most of the interventions. While the projects proved meaningful for those involved with them and exposed to them, participants represented just a sliver of the population in most communities. Further, they often included individuals who were already open to ideas of arts-based activism, or themes of resistance, resilience and/ or reconciliation, as well as the potential to transform and disseminate these through remediation. This should not discount the impact made on individuals who were involved in the various interventions. As is in the case of musically mediated empathy, questions remain about if or how to 'scale up' some of the benefits that we observed. Should we strive to include more people in dialogues and participatory theatre productions or is the intimacy of the current spaces what makes critical discussions and community building possible? Should such initiatives be replicated or do they risk losing their grassroots nature and risk becoming too standardized to be meaningful? These are ongoing questions for both theorists and practitioners.

Another limitation for resistance efforts was the broader political realities. In Lebanon, for example, very little has changed despite the mass mobilization of communities across the country during the 2019–20 revolution. Likewise, in Palestine, the conflict continues to elude any kind of progress or resolution, and the occupation in the West Bank becomes more entrenched every year. More than twenty years past the signing of the Good Friday Agreement, many parts of society in Northern Ireland remain divided – divisions that are stoked at times by new political tensions. One benefit of arts-based practices is that they are adaptable to a changing political atmosphere, although models designed at one point in time often are no longer appropriate at a later point, while a production or recording may lose resonance over time. Moreover, the protracted nature of many of the conflicts we studied can risk inducing activism fatigue and discourage resistance over time. Some may also question the feasibility or desirability of arts-based interventions in times of active conflict, especially if overlaid with a humanitarian crisis.

Finally, while we interrogate the interrelated nature of resistance, resilience, reconciliation and remediation in this book, the complexity of their interaction also can create challenges for practitioners. Should one of these elements be prioritized over others? How should partners manage the priorities of different participants and community members? What happens if the emphasis on resilience replaces or neutralizes resistance (Keelan and Browne 2020)? How might remediation weaken the impact of locally situated gestures of resistance or resilience? What happens if resistance or resilience manifest as forms of stubbornness that impede reconciliation? These are thorny issues that must be worked out on the ground, but we maintain that participatory arts-based practices can provide at least some communities with creative strategies for challenging the status quo. The nature and form of such interventions will look different in every context and will change and adapt over time. We hope this book will encourage others to further examine the role of sonic practices in shaping resistance, resilience, reconciliation and remediation in conflict and in post-conflict societies.

Soundscapes for change

The final 'red thread' of this volume is the combination, cognition and interpretation of sound and scoundscapes. In relation to the case studies presented throughout, soundscape operates as connecting tissue which contracts and expands depending on how it relates to practices such as music, theatre and sound arts. Soundscape encompasses the sonic realities of the participatory interventions that we discuss but also those of the everyday – most critically those realities which rely on the sonic for navigating complex and dangerous situations (Chapter 6). As a term, 'soundscape' is inherently relational in that it sits between acoustic phenomena and individual or group experience. This relational quality is critical in helping avoid caricaturing or oversimplifying the listening experience. Thinking relationally about sound moves us away from a situation of causality in which sound x produces response y towards a more holistic understanding of the various conditions that affect the listening situation. Soundscape then becomes not a phenomenon to be described or analysed but rather one that is produced. We occupy a position of creative agency as both listeners and makers of our own soundscape.

The International Standards Organization definition of soundscape reads as an 'acoustic environment as perceived or experienced and/or understood by a person or people in context' (ISO 2014). The importance of context and emphasis on relation in the Standard (e.g. between sources and environment, environment and perception, perception and interpretation and response etc.) highlights the risks in talking about sound as an object in a Schaefferian sense if we are concerned with the societal impact and potential of sound. Instead, a focus on perception, experience and understanding shifts the emphasis from sound source to an individual or shared sonic meaning emerging through the relation of sound and its perceivers. This is reflected in UNESCO's statement in 'The Importance of Sound in Today's World: Promoting Best Practices' that 'the sound environment is a mirror and a gateway to the world; it reflects and shapes our individual and collective behaviour, and our productivity and capacity to live in harmony together' (2017: 1). Both the ISO standard and UNESCO's claim position sounds as mechanisms for social and political change.

Recalling the sonic potential to drive change is crucial for a full understanding of sounding (post-)conflict. Likewise, social and political change implies a reimagining of our soundscapes and our relations with others. Artistic interventions can be a space for just such a reimagining: a space to ask what if? to take risks and let sonic imagination fuel our interconnectedness. A shared listening to a reimagined soundscape is a powerful tool for momentary displacement from the here and now, while sound's capacity for unfolding space and time creates opportunities for questioning inevitability and listening otherwise. To reimagine our soundscape is to reimagine our being in the world.

Resistance, resilience, reconciliation and remediation have served as frameworks for discussing sonic practices throughout this book, but we close with an invitation to a sonic reimagining that draws together listening, engagement, storytelling and sound making. This is imbricated with questions of ethics, aesthetics and politics, but as simultaneously embodied listeners and sound makers, we are responsible for (and response*able* to) our soundscapes. A sonic reimagining brings these questions to the fore in acts of hope and steps for change.

References

Aboultaif, E. W., and P. Tabar (2019), 'National versus Communal Memory in Lebanon', *Nationalism and Ethnic Politics*, 25 (1) 97–114.

Abrahams, D., and J. Lily (2015), *Circular Sounds*. Available online: https://darrenabrahams.com/tools/ (accessed 23 July 2020).

Adnan, S. (2007), 'Departures from Everyday Resistance and Flexible Strategies of Domination: The Making and Unmaking of a Poor Peasant Mobilization in Bangladesh', *Journal of Agrarian Change*, 7: 183–224.

Aissa, C. (2017), *Musicians without Borders: A Case Study Written by LEYMN Apprentice Cina Aissa*, London: Sound Connections LEYMN. Available online: https://www.sound-connections.org.uk/wp-content/uploads/SC-Musicians-without-borders.pdf (accessed 23 February 2021).

Alan in Belfast Blog (2018), 'So I Can Breathe This Air: A Study of Belonging, Identity and Home While Walking across Belfast', 5 August 2018. Available online: http://alaninbelfast.blogspot.com/2018/08/so-i-can-breathe-this-air-study-of.html (accessed 16 September 2022).

Amnesty International (2019), 'Lebanon Protests Explained'. Available online: https://www.amnesty.org/en/latest/news/2019/11/lebanon-protests-explained/ (accessed 16 September 2022).

Attali, J. (1985), *Noise: The Political Economy of Music*, trans. B. Massumi, Minneapolis: University of Minnesota Press.

Augoyard, J.-F., and H. Torgue (2005), *Sonic Experience: A Guide to Everyday Sounds*, trans. A. McCartney and D. Paquette, Montreal: McGill-Queen's University Press.

Austin, D. (2002), 'The Wounded Healer: The Voice of Trauma: A Wounded Healer's Perspective', in J. Sutton (ed.), *Music, Music Therapy and Trauma: International Perspectives*, 231–59, London: Jessica Kingsley.

Azar, R. (2020), 'A Response to "Listening with Displacement"', *Migration and Society*, 3 (1): 310–5. doi: https://doi.org/10.3167/arms.2020.030129

Baaz, M., M. Lilja, M. Schulz and S. Vinthagen (2016), 'Defining and Analyzing "Resistance": Possible Entrances to the Study of Subversive Practices', *Alternatives: Global, Local, Political*, 41 (3) (August): 137–53.

Baker, G. (2014), *El Sistema: Orchestrating Venezuela's Youth*, Oxford: Oxford University Press.

Bang, A. H. (2016), 'The Restorative and Transformative Power of the Arts in Conflict Resolution', *Journal of Transformative Education*, 14 (4): 355–76.

Barenboim, D. (2006), 'Reith Lecture 4: Meeting in music' [BBC Reith Lectures 2006, Transcript]. Available online: http://downloads.bbc.co.uk/rmhttp/radio4/transcripts/20060428_reith.pdf (accessed 20 September 2019).

Bartels, D. (2019), 'Regarding Young Musicians as Ethical and Aesthetic Practitioners: A New Reading of *Phronēsis*', *Philosophy of Music Education Review*, 27 (1): 37–50.

Barthes, R. ([1972] 1977), 'The Grain of the Voice', in *Image – Music – Text*, trans. S. Heath, 179–89, London: Fontana Press.

Bekerman, Z., and M. Zembylas (2012), *Teaching Contested Narratives: Identity, Memory, and Reconciliation in Peace Education and Beyond*, Cambridge: Cambridge University Press.

The Belfast Agreement: Agreement Reached in the Multi-Party Negotiations, 10 April 1998 (also known as the Good Friday Agreement). Available online: www.gov.uk/government/publications/the-belfast-agreement (accessed 7 September 2022).

Bell, L. A. (2019), *Storytelling for Social Justice: Connecting Narrative and the Arts in Antiracist Teaching*, 2nd edn, London: Routledge.

Bensimon, M., D. Amir and Y. Wolf (2008), 'Drumming through Trauma: Music Therapy with Post-traumatic Soldiers', *Arts in Psychotherapy*, 35: 34–48.

Bishop, C. (2012), *Artificial Hells: Participatory Art and the Politics of Spectatorship*, London: Verso Books.

Bloomfield, D. (2006), *On Good Terms: Clarifying Reconciliation*, Berghof Report No. 14, Berlin: Berghof Research Center for Constructive Conflict Management.

Bloomfield, D., T. Barnes and L. Huyse, eds (2003), *Reconciliation after Violent Conflict: A Handbook*, Stockholm: International Institute for Democracy and Electoral Assistance.

Bogart, A. (2015), 'The Role of Storytelling in the Theatre of the Twenty-First Century', Address to the Humana Festival of New American Plays, Louisville, 13 May. Transcript available online: https://howlround.com/role-storytelling-theatre-twenty-first-century (accessed 18 November 2021).

Bolter, J. D., and R. Grusin (1999), *Remediation: Understanding New Media*, Cambridge: MIT Press.

Born, G. (2010a), 'For a Relational Musicology: Music and Interdisciplinarity, Beyond the Practice Turn', *Journal of the Royal Musical Association*, 135 (2): 205–43.

Born, G. (2010b), 'Listening, Mediation, Event: Anthropological and Sociological Perspectives', *Journal of the Royal Musical Association*, 135 (S1): 79–89. doi: https://doi.org/10.1080/02690400903414855.

Bourriaud, N. (2002), *Relational Aesthetics*, trans. S. Pleasance and F. Woods, Dijon: Les presses du réel.

Bowman, W. (2016), 'Artistry, Ethics, and Citizenship', in D. J. Elliott, M. Silverman and W. Bowman (eds), *Artistic Citizenship: Artistry, Social Responsibility, and Ethical Praxis*, 59–80, New York: Oxford University Press.

Brenneis, D. (1987), 'Performing Passions, Aesthetics and Politics in an Occasionally Egalitarian Community', *American Ethnologist*, 14 (2): 236–50.

Brewer, J., B. Hayes and F. Teeney (2018), *The Sociology of Compromise after Conflict*. Palgrave Studies in Compromise after Conflict, London: Palgrave Macmillan.

Brown, R. (2010), *Sound: A Reader in Theatre Practice*. Basingstoke: Palgrave Macmillan.

Brown, R. (2020), *Sound Effect: The Theatre We Hear*. London: Bloomsbury Methuen Drama.

Buffalo String Works (2021a), Concert Programme for *They Would Only Walk*. 20 March 2021.

Bull, M. (2021), 'Introduction to Part 1: Sounds Inscribed onto the Face: Rethinking Sonic Connections through Time, Space, and Cognition', in M. Bull and M. Cobussen (eds), *The Bloomsbury Handbook of Sonic Methodologies*, 17–34, London: Bloomsbury Handbooks.

Bull, M., and M. Cobussen, eds (2021), *The Bloomsbury Handbook of Sonic Methodologies*, London: Bloomsbury Handbooks.

Burnard, P., V. Ross, L. Hassler and L. Murphy (2018), 'Translating Intercultural Creativities in Community Music', in B.-L. Bartleet and L. Higgins (eds), *The Oxford Handbook of Community Music*, 229–42. Oxford: Oxford University Press.

Butler, J. (2016), 'Rethinking Vulnerability and Resistance', in J. Butler, Z. Gambetti and L. Sabsay (eds), *Vulnerability in Resistance*, 12–27. Durham, NC: Duke University Press.

Cao, E. L., C. D. Blinderman and I. Cross (2021), 'Reconsidering Empathy: An Interpersonal Approach for Participatory Arts in the Medical Humanities', *Journal of the Medical Humanities*, 42 (December): 627–40. doi: https://doi.org/10.1007/s10912-021-09701-6.

Cavarero, A. (2000), *Relating Narratives: Storytelling and Subjecthood*, trans. P. A. Kottman, London: Routledge.

Cavarero, A. (2005), *For More than One Voice: Toward a Philosophy of Vocal Expression*, trans. P. A. Kottman, Stanford, CA: Stanford University Press.

Cavarero, A., K. Thomaidis and I. Pinna (2018), 'Towards a Hopeful Plurality of Democracy: An Interview on Vocal Ontology with Adriana Cavarero', *Journal of Interdisciplinary Voice Studies*, 3 (1): 81–93. doi: https://doi.org/10.1386/jivs.3.1.81_1

Chilton, G., and P. Leavy (2014), 'Arts-Based Research Practice: Merging Social Research and the Creative Arts', in P. Leavy (ed.), *The Oxford Handbook of Qualitative Research*, 403–22, New York: Oxford University Press.

Chion, M. (1994), *Audio-Vision: Sound on Screen*, trans. C. Gorbman, New York: Columbia University Press.

Clandinin, D. J., and F. M. Connelly (2000), *Narrative Inquiry: Experience and Story in Qualitative Research*, San Francisco: Jossey Bass.

Clifford, J. (1986), 'Introduction: Partial Truths', in J. Clifford and G. E. Marcus (eds), *Writing Culture: The Poetics and Politics of Ethnography*, 1–26, Berkeley: University of California Press.

Clinton, W. J. (1995), *Speech by the President of USA to the People of Derry*, 30 November. Transcript available online: https://cain.ulster.ac.uk/events/peace/docs/pres2.htm (accessed 17 November 2021).

Cobussen, M. (2014), 'Steps to an Ecology of Improvisation', in F. Schroeder and M. O hAodha (eds), *Soundweaving: Writings on Improvisation*, 31–45, Cambridge: Cambridge Scholars.

Cohen, C. (2006), 'Creative Approaches to Reconciliation', in M. Fitzduff and C. Stout (eds), *The Psychology of Resolving Global Conflicts: From War to Peace*, vol. 3: *Interventions*, 69–102, Westport, CT: Greenwood.

Cohen, C. (2008), 'Music: A Universal Language?', in O. Urbain (ed.), *Music and Conflict Transformation: Harmonies and Dissonances in Geopolitics*, 26–39, London: I.B. Tauris.

Cohen, C., R. G. Varea and P. Walker (2011), *Acting Together: Performance and the Creative Transformation of Conflict* (Vols. 1–2), Oakland: New Village Press.

Collard, R. (2019), 'Untouchable No More: Hezbollah's Fading Reputation', *Foreign Policy*. Available online: https://foreignpolicy.com/2019/11/27/lebanon-protests-hezbollah-fading-reputation/ (accessed 8 September 2022).

Colvin, C. (2018), *Traumatic Storytelling and Memory in Post-Apartheid South Africa: Performing Signs of Injury*, London: Routledge.

Connor, S. (1997), 'The Modern Auditory I', in R. Porter (ed.), *Rewriting the Self: Histories from the Renaissance to the Present*, 203–23, London: Routledge.

Connor, S. (2010), *Steven Connor: Auscultations (Listening In) (Sonic Acts XIII, 2010)*. Available online: https://vimeo.com/12701700 (accessed 15 September 2010).

Cook, N. (2012), 'Anatomy of the Encounter', in S. Hawkins (ed.), *Critical Musicological Reflections: Essays in Honour of Derek B Scott*, 193–208, Farnham, UK: Ashgate.

Corbin, A. (1998), *Village Bells: Sound and Meaning in the 19th-century French Countryside*, trans. M. Thom, New York: Columbia University Press.

Croce, N. D. (2017), '"You Can Hear Them before You See Them": Listening through Belfast Segregated Neighborhoods', *Journal of Sonic Studies*, 14. Available online: https://www.researchgate.net/publication/322553931_You_can_hear_them_before_you_see_them_Listening_through_Belfast_segregated_neighborhoods/link/5a5f871ba6fdcc21f4857aa5/download (accessed 16 September 2022).

Csikszentmihályi, M. (1990), *Flow: The Psychology of Optimal Experience*, New York: Harper and Row.

Daughtry, M. (2015), *Listening to War: Sound, Music, Trauma and Survival in Wartime Iraq*, Oxford: Oxford University Press.

Davis, C., and H. Meretoja (2017), *Storytelling and Ethics: Literature, Visual Arts, and the Power of Narrative*, London: Routledge.

De Nora, T. (2000), *Music in Everyday Life*, Cambridge: Cambridge University Press.

Derrida, J. (1976), *Of Grammatology*, trans. G. C. Spivak, Baltimore, MD: Johns Hopkins University Press.

d'Estree, T. P. (2005), 'The Role of "Voice" in Intergroup Conflict De-escalation and Resolution', in M. Fitzduff and C. E. Stout (eds), *The Psychology of Resolving Global Conflicts: From War to Peace*, vol. 3. 103–22, Westport, CT: Praeger Security International.

Diamond, B. (2015), 'The Doubleness of Sound in Canada's Indian Residential Schools', in V. L. Levine and P. V. Bohlman (eds), *This Thing Called Music: Essays in Honor of Bruno Nettl*, 267–79, Lanham, MD: Rowman and Littlefield.

Dillon. S. (2007), 'Maybe We Can Find Some Common Ground: Indigenous Perspectives, a Music Teacher's Story', *Australian Journal of Indigenous Education*, 36 Supplement: 59–65.

Dillon, S. (2011), '*Music Is a Wordless Knowing of Others: Resilience in Virtual Ensembles*', in A. Brader (ed.), *Songs of Resilience*, 239–55, Cambridge: Cambridge Scholars.

Dirkx, J. M. (2006), 'Engaging Emotions in Adult Learning: A Jungian Perspective on Emotion and Transformative Learning', *New Directions for Adult and Continuing Education*, 109 (Spring): 15–26. doi: https://doi.org/10.1002/ace.204.

Dolan, J. (2005), *Utopia in Performance. Finding Hope at the Theatre*, Ann Arbor: University of Michigan.

Donaghey, J., and F. Magowan (2021), 'Emotion Curves: Creativity and Methodological "Fit" or "Commensurability"', *International Review of Qualitative Research*, 15 (1): 1–18. doi: https://doi.org/10.1177/19408447211002768

Dowling, M. (2008), 'Fiddling for Outcomes: Traditional Music, Social Capital and Arts Policy in Northern Ireland', *International Journal of Cultural Policy*, 14 (2): 179–94.

Duncombe, S. (2007), '(From) Cultural Resistance to Community Development', *Community Development Journal*, 42 (4): 490–500.

Eidsheim, N. S. (2015), *Sensing Sound: Singing and Listening as Vibrational Practice*, Durham, NC: Duke University Press.

Emberley, J. V. (2014), *The Testimonial Uncanny: Indigenous Storytelling, Knowledge, and Reparative Practices*, Albany: State University of New York Press.

Emery, D. B. (1993), 'Self, Creativity, Political Resistance', *Political Psychology*, 14 (2): 347–62.

Empathy for Peace (2019), 'Empathy: An Invaluable Natural Resource for Peace'. Available online: https://www.empathy-for-peace.org/ (accessed 4 January 2022).

English, L. (2017), 'Relational Listening: A Politics of Perception', *Contemporary Music Review*, 36 (3): 127–42. doi: https://doi.org/10.1080/07494467.2017.1395141.

Enria, L. (2016), 'Co-producing Knowledge through Participatory Theatre: Reflections on Ethnography, Empathy and Power', *Qualitative Research*, 16 (3): 319–29.

Erikson, E. H. (1963), *Childhood and Society*, 2nd edn, New York: Norton.

Erll, A., and A. Rigney, eds (2009), *Mediation, Remediation and the Dynamics of Cultural Memory*, Berlin: Walter de Gruyter.

Erlmann, V. (2004), 'But What of the Ethnographic Ear? Anthropology, Sound, and the Senses', in *Hearing Cultures: Essays on Sound, Listening and Modernity*, 1–20, Abingdon: Routledge.

Feld, S. (1982), *Sound and Sentiment: Birds, Weeping and Poetic Song in Kaluli Expression*, Durham, NC: Duke University Press.

Feld, S. (2015), 'Acoustemology', in D. Novak and M. Sakakeeny (eds), *Keywords in Sound*, 12–21. Durham, NC: Duke University Press.

Feld, S., and D. Brenneis (2004), 'Doing Anthropology in Sound', *American Ethnologist*, 31 (4): 461–74.

Fiddian-Qasmiyeh, E. (2020), 'Refuge in a Moving World: Refugee and Migrant Journeys across Disciplines', in E. Fiddian-Qasmiyeh (ed.), *Refuge in a Moving World: Tracing Refugee and Migrant Journeys across Disciplines*, 1–20, London: University College London Press.

Fischer-Lichte, E. (2008), *The Transformative Power of Performance: A New Aesthetics*, trans. S. I. Jain, Oxford/New York: Routledge [Original: *Ästhetik des Performativen* (2004)].

Fitzpatrick, L. (2009), 'Utopian Performative in Post-ceasefire Northern Irish Theatre', in L. Fitzpatrick (ed.), *Performing Violence in Contemporary Ireland*, 175–188, Dublin: Carysfort Press.

Foley, E. (2020), *Music: Its Theologies and Spiritualities*, Basel: MDPI. doi: https://doi.org/10.3390/books978-3-03943-594-4.

Foster, V. (2012), 'The Pleasure Principle: Employing Arts-based Methods in Social Work Research', *European Journal of Social Work*, 15 (4): 532–45.

Foucault, M. (1978), *The History of Sexuality: An Introduction*, New York: Random House.

Gallagher, M., and J. Prior (2014), 'Sonic Geographies: Exploring Phonographic Methods', *Progress in Human Geography*, 38 (2): 267–84.

Ganz, M. (2007), *What Is Public Narrative?* Cambridge, MA: Harvard University, Kennedy School of Government.

Gershon, W. (2018), 'Resounding Science: A Sonic Ethnography of an Urban Fifth Grade Classroom', *Journal of Sonic Studies*, no. 04 (29 June). Available online: https://www.researchcatalogue.net/view/290395/290396 (accessed 8 September 2022).

Gibbons, R. C. (2001), 'Trust in Social Structures: Hobbes and Coase meet Repeated Games', in K. S. Cook (ed.), *Trust in Society*, 322–53, New York: Russel Sage Foundation.

Giddens, A. (1990), *The Consequences of Modernity*, Stanford, CA: Stanford University Press.

Gopinath, S. (2009), 'The Problem of the Political in Steve Reich's *Come Out*', in R. Adlington (ed.), *Sound Commitments: Avant-Garde Music and the Sixties*, 121–40, Oxford: Oxford University Press.

Gordon, T. (2019), 'Spotlight on Mary Kouyoumdjian', *PSNY*, 17 September, Schott Music Group. Available online: https://www.eamdc.com/psny/blog/spotlight-on-mary-kouyoumdjian/ (accessed 24 September 2021).

Grant, M. J., R. Möllemann, I. Morlandstö, S. C. Münz and C. Nuxoll (2010), 'Music and Conflict: Interdisciplinary Perspectives', *Interdisciplinary Science Reviews*, 35 (2): 183–98.

Green, L. (2008), *Music, Informal Learning and the School*, London: Ashgate.

Greene, J., V. Caracelli and W. Graham (1989), 'Toward a Conceptual Framework for Mixed-Method Evaluation Designs', *Educational Evaluation and Policy Analysis*, 11 (3): 255–74.

Grønseth, A. S. (2010), 'Sharing Experiences with Tamil Refugees in Norway', in A. S. Grønseth and D. Davis (eds), *Mutuality and Empathy: Self and Other in the Ethnographic Encounter*, 143–61, Wantage, UK: Sean Kingston.

Haddad, S. (2009), 'Lebanon: From Consociationalism to Conciliation'. *Nationalism and Ethnic Politics*, 15 (3–4): 398–416.

Haile, S. (2020), 'Voices to Be Heard? Reflections on Refugees, Strategic Invisibility and the Politics of the Voice', in E. Fiddian-Qasmiyeh (ed.), *Refuge in a Moving World*, 32–40, London: UCL Press.

Head, N. (2012), 'Transforming Conflict: Trust, Empathy, and Dialogue', *International Journal of Peace Studies*, 17 (2): 33–55.

Heaney, S. (1990), *The Cure at Troy*, London: Faber.

Heath, S., L. Chapman and the Morgan Centre Sketchers (2018), 'Observational Sketching as Method', *International Journal of Social Research Methodology*, 21 (6): 713–28. doi: https://doi.org/10.1080/13645579.2018.1484990

Hegarty, P. (2007), *Noise/Music: A History*. New York: Continuum.

Heidemann, B. (2016), *Post-Agreement Northern Irish Literature: Lost in a Liminal Space?* Basingstoke: Palgrave Macmillian.

Heilbronner, O. (2016), 'Music and Protest: The Case of the 1960s and its Long Shadow', *Journal of Contemporary History*, 51 (3): 688–700.

Helguera, P. (2011), *Education for Socially Engaged Art: A Materials and Techniques Handbook*. New York: Jorge Pinto Books.

Higgins, L. (2012), *Community Music in Theory and in Practice*. Oxford: Oxford University Press.

Hirschkind, C. (2006), *The Ethical Soundscape: Cassette Sermons and Islamic Counterpublics*. New York: Columbia University Press.

Hollander, J. A., and R. L. Einwohner (2004), 'Conceptualizing Resistance', *Sociological Forum*, 19 (4): 533–54.

Howell, G. (2018), 'Harmony', *Music & Arts in Action*, 6 (2): 45–58.

Howes, D. (2005) ed., *Empire of the Senses: The Sensual Culture Reader*, Oxford: Berg.

Howes, D., and C. Classen (2013), *Ways of Sensing: Understanding the Senses in Society* 1st edn, London: Routledge.

Hughes, J. (2007), 'Mediating and Moderating Effects of Inter-group Contact: Case Studies from Bilingual/Bi-national Schools in Israel', *Journal of Ethnic and Migration Studies*, 33 (3): 419–37.

Ihde, D. (2007), *Listening and Voice: Phenomenologies of Sound*, 2nd edn, Albany: State University of New York Press.

ISO [International Standards Organization] (2014), 'Acoustics: Soundscape', British Standards, International Standards Organization 12913-1.

James, R. (2012), 'Affective Resonance: On the Uses and Abuses of Music in and for Philosophy', *Phaenex: Journal of Existential and Phenomenological Theory and Culture*, 7 (2): 59–95.

James, R. (2019), *The Sonic Episteme: Acoustic Resonance, Neoliberalism, and Biopolitics*, Durham, NC: Duke University Press.

Jarry, A. (1965), *Selected Works of Alfred Jarry*, R. Shattuck and S. W. Taylor (eds), New York: Grove Press.

Johansson, A., and S. Vinthagen (2016), 'Dimensions of Everyday Resistance: An Analytical Framework', *Critical Sociology*, 42 (3): 417–35.

Johansson, A., and S. Vinthagen (2020), *Conceptualizing 'Everyday Resistance': A Transdisciplinary Approach*, New York: Routledge.

Jorgensen, D. L. (2015), 'Participant Observation', *Emerging Trends in the Social and Behavioral Sciences*, Wiley Online Library (15 May). doi: https://doi.org/10.1002/9781118900772.etrds0247

Josephides, L. (2010), 'Speaking-with and Feeling-with: The Phenomenology of Knowing the Other', in A. S. Grønseth and D. Davis (eds), *Mutuality and Empathy: Self and Other in the Ethnographic Encounter*, 161–76, Wantage UK: Sean Kingston.

Juntunen, M.-L., S. Karlsen, A. Kuoppamäki, T. Laes and S. Muhonen (2014), 'Envisioning Imaginary Spaces for Musicking: Equipping Students for Leaping into the Unexplored', *Music Education Research*, 16 (3): 251–66.

Keelan, E., and B. Browne (2020), 'Problematising Resilience: Development Practice and the Case of Palestine', *Development Practice*, 30 (4): 459–71.

Kendrick, L., and D. Roesner, eds (2011), *Theatre Noise*, Newcastle upon Tyne: Cambridge Scholars.

Kersenboom, S. (1995), *Word, Sound, Image: The Life of the Tamil Text*, Oxford: Berg.

Khader, S. J. (2018), ''Victims' Stories and the Postcolonial Politics of Empathy', *Metaphilosophy*, 49 (1–2): 13–26.

Kim-Cohen, S. (2009), *In the Blink of an Ear: Towards a Non-Cochlear Sonic Art*, New York: Continuum.

King, E., and C. Waddington eds (2017), *Music and Empathy*, New York: Routledge.

Klein, J. (2017), 'What Is Artistic Research?', *Journal of Artistic Research*, jar-online.net, doi: https://doi.org/10.22501/jarnet.0004

Korum, S. (2020), 'The Sound of Reconciliation? Musical and Sociocultural Harmony in the Sri Lanka Norway Music Cooperation (2009–2018)', *Asian European Music Research Journal*, 5 (June): 51–65. doi: https://doi.org/10.30819/aemr.5-7

Kouyoumdjian, M. (2016), *Paper Pianos: 'You Are not a Kid'*, 9′08″. Recording by Alarm Will Sound. Available online: https://soundcloud.com/alarm-will-sound/you-are-not-a-kid-from-paper-pianos (accessed 24 September 2021).

Kouyoumdjian, M. (2020), *They Would Only Walk*, virtual premiere 20 March 2021 by Buffalo String Works and Buffalo Chamber Players, Buffalo, NY.

Kouyoumdjian, M. (2021), 'Creating with Ghosts: Identity and Artistic Purpose in Armenian Diaspora', DMA thesis, Columbia University. doi: https://doi.org/10.7916/d8-4fqv-ch76

Krog, A. (1999), *Country of My Skull*, London: Jonathan Cape.

LaBelle, B. (2010), *Acoustic Territories: Sound Culture and Everyday Life*, London: Bloomsbury Academic.

LaBelle, B. (2018), *Sonic Agency: Sound and Emergent Forms of Resistance*, London: Goldsmiths Press, University of London.

LaBelle, B. (2021), *Acoustic Justice: Listening, Performativity, and the Work of Reorientation*, London: Bloomsbury.

Lacy, S., ed. (1995), *Mapping the Terrain: New Genre Public Art*, Seattle, WA: Bay Press.

Lakoff, G., and M. Johnson (2003), *Metaphors We Live By*. With a new Afterword, Chicago, IL: University of Chicago Press.

Larkin, C. (2010), 'Beyond the War? The Lebanese Postmemory Experience', *International Journal Middle East Studies*, 42 (4): 615–35.

Launchbury, C., N. Tamraz, R. Celestin and E. DalMolin (2014), 'War, Memory, Amnesia: Postwar Lebanon', *Contemporary French and Francophone Studies*, 18 (5): 457–61.

Laurence, F. (2008), 'Music and Empathy', in O. Urbain (ed.), *Music and Conflict Transformation: Harmonies and Dissonance in Geopolitics*, 13–25, London: I.B. Tauris.

Laurence, F. (2017), 'Prologue: Revisiting the Problem of Empathy', in E. King and C. Waddington (eds), *Music and Empathy*, 11–35, New York: Routledge.

Lederach, J. P. (2001), 'Civil Society and Reconciliation', in C. A. Crocker, F. O. Hampson and P. Aall (eds), *Turbulent Peace: The Challenges of Managing International Conflict*, 841–54. Washington, DC: United States Institute of Peace.

Lederach, J. P. (2005), *The Moral Imagination: The Art and Soul of Building Peace*, Oxford: Oxford University Press.

Lederach, J. P. (2011), 'Foreword: Acting Together on the World Stage', in C. Cohen, R. G. Varrea and P. O. Walker (eds), *Acting Together: Performance and the Creative Transformation of Conflict*, ix–xii, Oakland, CA: New Village Press.

Leeuwen, T. van. (1998), 'Music and Ideology: Notes toward a Sociosemiotics of Mass Media Music', *Popular Music and Society*, 22 (4) (1 December): 25–54. doi: https://doi.org/10.1080/03007769808591717

Lefebvre, H. (2004), *Rhythmanalysis: Space, Time and Everyday Life*, London: Continuum. doi: http://dx.doi.org/10.5040/9781472547385

Lehner, S. (2011), 'Post-Conflict Masculinities: *Filiative* Reconciliation in *Five Minutes of Heaven* and David Park's *The Truth Commissioner*', in C. Magennis and R. Mullins (eds), *Irish Masculinities: Critical Reflections on Literature and Culture*, 65–76, Dublin: Irish Academic Press.

Lehner, S. (2020), 'Nation: Reconciliation and the Politics of Friendship in Post-Troubles Literature', in P. Reynolds (ed.), *The New Irish Studies: Twenty-First Century Critical Revisions*, 47–62, Cambridge: Cambridge University Press.

Leonard, J. A. (2001), *Theatre Sound*. New York: Routledge.

Levine, P. A. (2010), *In an Unspoken Voice: How the Body Releases Trauma and Restores Goodness*, Berkeley, CA: North Atlantic Books.

Magowan, F., and J. Donaghey (2018), 'Musicians without Borders Music Bridge and Training of Trainers 2017 Report'. Available online: https://www.qub.ac.uk/research-centres/SoundingConflict/FileStore/Filetoupload,885080,en.pdf (accessed 17 November 2021).

Magowan, F., and H. Donnan (2019), 'Introduction: Sounding and Performing Resistance and Resilience', *Music and Arts in Action*, 7 (1): 1–10.

Manzini, E., and J. Till (2015), 'The Cultures of Resilience Base Text', in E. Manzini and J. Till (eds), *Cultures of Resilience: Ideas*, 9–12, London: Hato Press.

MAP Fund (2021), '2021 Grant Cycle Process and Reflections'. Available online: https://mapfundblog.org/2021-grant-cycle-process-and-reflections/ (accessed 24 September 2021).

Martinez, T. A. (1997), 'Popular Culture as Oppositional Culture: Rap as Resistance', *Sociological Perspectives*, 40 (2): 265–86.

Mauss, M. (1973), 'Techniques of the Body', *Economy and Society*, 2: 70–88.

Maxwell, J. A. (2005), *Qualitative Research Design: An Interactive Approach*, 2nd edn, Thousand Oaks, CA: Sage.

McAdam, D., Tarrow S. and Tilly, C. (2001), *Dynamics of Contention*, Cambridge: Cambridge University Press.

McEvoy, J. (2008), *The Politics of Northern Ireland* (Politics Study Guides), Edinburgh: Edinburgh University Press.

McKeown, L. (2014), *Those You Pass on the Street* (unpublished script).

McKeown, L. (2016), *Green and Blue* (unpublished script).

McMurray, P. (2021), 'Mobilizing Karbala: Sonic Remediations in Berlin Ashura Processions', *Ethnic and Racial Studies*, 44 (10): 1864–85. doi: 10.1080/01419870.2020.1846763

McRae, C. (2020), 'Performative Listening', in D. L. Worthington and G. D. Bodie (eds), *The Handbook of Listening*, 399–408, Hoboken, NJ: Wiley-Blackwell.

Middleton, I. (2018), Trust in Music: Musical Projects against Violence in Northern Colombia, PhD dissertation, University of Illinois, Urbana Champaign.

Mills, S. (2014), *Auditory Archaeology: Understanding Sound and Hearing in the Past*, Walnut Creek, CA: Left Coast Press.

Milton-Edwards, B. (2021), 'We Are The Ones That Are Impatient: Improvising Resistance and Resilience in Jordanian Hip Hop and Rap', in D. Fischlin and E. Porter (eds), *Sound Changes: Improvisation and Transcultural Difference*, 76–100, Ann Arbor: University of Michigan Press.

Mühlhoff, R. (2015), 'Affective resonance and social interaction', *Phenomenology and the Cognitive Sciences*, 14 (4): 1001–19.

Myers, D. (2016), *Victims' Stories and the Advancements of Human Rights*, Oxford: Oxford University Press.

Myers M. (2008), 'Situations for living: Performing emplacement', *Research in Drama Education*, 13 (2): 171–80.

Myers, M. (2010), '"Walk with Me, Talk with Me": The Art of Conversive Wayfinding', *Visual Studies*, 25 (1): 59–68.

Myers, M. (2011), 'Vocal Landscaping: The Theatre of Sound in Audiowalks', in L. Kendrick and D. Roesner (eds), *Theatre Noise*, 70–81, Newcastle upon Tyne: Cambridge Scholars.

Myers, M., D. Watkins and R. Sobey (2016), 'Conversive Theatres: Performances with/in Social Media', *Performance Paradigm*, 12: 82–98.

Nancy, J. -L. (2007), *Listening*, trans. C. Mandell, New York: Fordham University Press.

Ng-A-Fook, N., and K. R. Llewellyn, eds (2019), *Oral History Education, Public Schooling, and Social Justice: Troubling Cultures of Reconciliation*, London: Routledge.

Nielsen T. R. (2015), 'Theatrical Complicity as a Medium of Emancipation', *Nordic Theatre Studies*, 27 (2): 48–59.

Noddings, N. (2002), *Starting at Home: Caring and Social Policy*, Berkeley: University of California Press.

Noddings, N. (2005), *The Challenge to Care in Schools: An Alternative Approach to Education*, New York: Teachers College Press.

Norman, J. M. (2009), 'Creative Activism: Youth Media in Palestine', *Middle East Journal of Culture and Communication* (2): 251–74.

Norman, J. M. (2010), 'The Activist and the Olive Tree: Nonviolent Resistance in the Second Intifada', Washington, DC: School of International Service, American University.

Norman, J. M. (2020), 'Beyond Hunger Strikes: The Palestinian Prisoners' Movement and Everyday Resistance', *Journal of Resistance Studies*, 6 (1): 40–68.

Novak, D., and M. Sakakeeny, eds (2015), *Keywords in Sound*, Durham, NC: Duke University Press.

Obert, K. (2016), 'What We Talk About When We Talk About Intimacy', *Emotion, Space and Society*, 21 (November): 25–32.

Odena, O., and L. Cabrera (2006), 'Dramatising the Score: An Action-Research Investigation of the use of Mozart's Magic Flute as Performance Guide for his Clarinet Concerto', in M. Baroni, A. R. Addessi, R. Caterina and M. Costa (eds), Proceedings of the 9th International Conference on Music Perception and Cognition, CD, Bologna, Italy: Society for Music Perception and Cognition and European Society for the Cognitive Sciences of Music.

Ó hÍr, L., and L. Strange. (2021), 'Tiocfaidh Ár Lá, Get the Brits out, Lad: Masculinity and Nationalism in Irish-Language Rap Videos', *Social Semiotics*, 31 (3): 466–88. doi: https://doi.org/10.1080/10350330.2021.1930856 (accessed 16 September 2022).

Oliveira, N. D., N. Oxley, M. Petry and M. Archer. (1994), *Installation Art*. Washington, DC: Smithsonian Institution Press.

Onion, R. (2014), 'The Colorful Quilt Squares Chilean Women Used to Tell the Story of Life Under Pinochet', *Slate*, 10 September. Available online: https://slate.com/human-interest/2014/09/history-of-quilting-arpilleras-made-by-chilean-women-to-protest-pinochet.html (accessed 17 November 2021).

O'Reilly, P. J. (2019), *Ubu the King* (unpublished rehearsal playscript).

O'Reilly, P. J. (2021), Email communication.

Palmer, V. J., C. Dowrick and J. M. Gunn (2014), 'Mandalas as a Visual Research Method for Understanding Primary Care for Depression', *International Journal of Social Research Methodology*, 17 (5): 527–41. doi: https://doi.org/10.1080/13645 579.2013.796764

Palombini, C. (2013), 'Funk Proibido', in L Avritzer, N. Bignotto, F. Filgueiras, J. Guimarães and H. Starling (eds), *Dimensões Políticas da Justiça*, 647–57, Rio de Janeiro: Civilização Brasileira. Available online: https://www.academia.edu/5268 654/_Funk_proibido_ (accessed 20 October 2021).

Parker, J. E. K. (2015), *Acoustic Jurisprudence*, New York: Oxford University Press.

Paulin, T. (1984), *The Riot Act*, London: Faber.

Pedwell, C. (2016), 'De-colonising Empathy: Thinking Affect Transnationally', *Samyutka: A Journal of Women's Studies*, Special Issue, 'Decolonising Theories of the Emotions', ed. S. Gunew, 16 (1): 27–49.

Peteet, J. (1996), 'The Writing on the Walls: The Graffiti of the Intifada', *Cultural Anthropology*, 11 (2): 139–59.

Peterson, E., and M. Sanouillet, (eds) (1989), *The Writings of Marcel Duchamp*, New York: Da Capo Press.

Phelan, M. (2016), 'From Troubles to Post-Conflict Theatre in Northern Ireland', in N. Grene and C. Morash (eds), *The Oxford Handbook of Modern Irish Theatre*, 372–88, Oxford: Oxford University Press.

Phillips-Hutton, A. (2018), 'Performing the South African Archive in *REwind: A Cantata for Voice, Tape, and Testimony*', *Twentieth-Century Music*, 15 (2): 187–209.

Phillips-Hutton, A. (2020), *Music Transforming Conflict*, Cambridge: Cambridge University Press.

Phillips-Hutton, A. (2021), 'Sonic Witnesses: Music, Testimony, and Truth', *Ethnomusicology Forum*, Special Issue: Music and Sound in Times of Violence, Displacement and Conflict, ed. F. M. Diaz and A. Wood, 30 (2): 266–82.

Pinch, T. (2016), 'The Sounds of Economic Exchange', in M. Bull and L. Back (eds), *The Auditory Culture Reader*, 2nd edn, 433–44, London: Bloomsbury.

Rabinger, M. (2009), *Directing the Documentary*, Abingdon, UK: Focal Press.

Rancière, J. ([1983] 2004a), *The Philosopher and His Poor*, ed. and intro. A. Parker; trans. J. Drury, C. Oster and A. Parker, Durham, NC: Duke University Press.

Rancière, J. ([2000] 2004b), *The Politics of Aesthetics*, trans. and intro. G. Rockhill; afterword S. Žižek, London: Continuum.

Reason, P. W., and H. Bradbury, (eds) (2007), *The Sage Handbook of Action Research: Participatory Inquiry and Practice*, 2nd edn, London: Sage.

Rebelo, P., and R. Cicchelli Velloso, (2018), 'Participatory Sonic Arts: Towards a Socially Engaged Art of Sound in the Everyday (The Som da Mare Project)', in S. Emmerson (ed.), *The Routledge Research Companion to Electronic Music: Reaching out with Technology*, 137–55, London: Routledge.

Reich, H. (2012), *The Art of Seeing: Investigating and Transforming Conflicts with Interactive Theatre*, Berlin: Berghof Foundation/Online Berghof Handbook for Conflict Transformation.

Reich, S. (1988), Programme note for *Different Trains* (1988), *Boosey & Hawkes*. Available online: www.boosey.com/cr/music/Steve-Reich-Different-Trains/2699 (accessed 24 September 2021).

Reich, S. (2002), *Writings on Music, 1965–2000*, ed. P. Hillier, Oxford: Oxford University Press.

Reynolds, D., and M. Reason, (eds) (2012), *Kinesthetic Empathy in Creative and Cultural Practices*, Chicago: University of Chicago Press.

Richards, S. (1995), 'In the Border Country: Greek Tragedy and Contemporary Irish Drama', in C. C. Barfoot and R. van den Doel (eds), *Ritual Remembering: History, Myth and Politics in Anglo-Irish Drama*, 191–200, Amsterdam: Rodopi.

Richards, S. (2008), 'Synge and the "Savage God"', *Études irlandaises*, 33 (2): 21–30.

Ricoeur, P. (1984), *Time and Narrative 1*, Chicago, IL: University of Chicago Press.

Ricoeur, P. (1999), 'Memory and Forgetting', in K. Richard and M. Dooley (eds), *Questioning Ethics: Contemporary Debates in Philosophy*, 5–11, London: Routledge.

Ricoeur, P. (2005), 'Memory, History, Forgiveness: A Dialogue Between Paul Ricoeur and Sorin Antohi', *Janus Head*, 8 (1): 14–25.

Robbins, J. (2007), 'Between Reproduction and Freedom: Morality, Value and Radical Cultural Changes', *Ethnos*, 72 (3): 293–314.

Rollefson, J. G. (2017), *Flip the Script: European Hip Hop and the Politics of Postcoloniality*, Chicago Studies in Ethnomusicology. Chicago, IL: University of Chicago Press. Available online: https://press.uchicago.edu/ucp/books/book/chicago/F/bo26955610.html (accessed 8 September 2022).

Rost, K. (2011), 'Intrusive Noises: The Performative Power of Theatre Sounds', in L. Kendrick and D. Roesner (eds), *Theatre Noise: The Sound of Performance*, 44–56, Newcastle upon Tyne: Cambridge Scholars.

Schaap, A. (2005), *Political Reconciliation*, London: Routledge.

Schafer, R. M. ([1977] 1994), *The Soundscape: Our Sonic Environment and the Tuning of the World*, Rochester, VT: Destiny Books.

Schutz, A. (1964), 'Making Music Together: A Study in Social Relationship', in A. Brodersen (ed.) *Alfred Schutz: Collected Papers II: Studies in Social Theory*, 159–78, The Hague: Nijhoff.

Schulze, H. (2021), 'Introduction: What Is an Anthropology of Sound?', in H. Schulze (ed.), *The Bloomsbury Handbook of the Anthropology of Sound*, 1–20, New York: Bloomsbury Academic.

Scott, J. C. (1985), *Weapons of the Weak*, New Haven, CT: Yale University Press.

Scott, J. C. (1990), *Domination and the Arts of Resistance: Hidden Transcripts*, New Haven, CT: Yale University Press.

Shaar, K. H. (2013), 'Post-Traumatic Stress Disorder in Adolescents in Lebanon as Wars Gained in Ferocity: A Systemic Review', *Journal of Public Health Research*, 2 (2): e17.

Shafiq (2017), *Interview with Julie Norman*, Bethlehem: Palestine

Shank, M., and L. Schirch (2008), 'Strategic Arts-Based Peacebuilding', *Peace and Change*, 33 (2): 217–42.

Sharp, G. (1973), *Power and Struggle: The Politics of Nonviolent Action*, Boston: Porter Sargent.

Shirlow, P. (2004), 'Northern Ireland: A Reminder from the Present', in C. Coulter and S. Coleman (eds), *The End of Irish History?*, 192–207, Manchester: Manchester University Press.

Skyllstad, K. (1997), 'Music in Conflict Management: A Multicultural Approach', *International Journal of Music Education*, 29: 73–80.

Skyllstad, K. (2008), 'Managing Conflict through Music: Educational Perspectives', in O. Urbain (ed.), *Music and Conflict Transformation: Harmonies and Dissonance in Geopolitics*, 172–83, London: I.B. Tauris.

Small, C. (1998), *Musicking: The Meanings of Performing and Listening*, Middletown, CT: Wesleyan University Press.

Smiley, S. (2015), 'Field Recording or Field Observation? Audio Meets Method in Qualitative Research', *The Qualitative Report*, 20 (16 November): 1812–22. doi: https://doi.org/10.46743/2160-3715/2015.2380

Smith, A., and A. Robinson (1996), *Education for Mutual Understanding: The Initial Statutory Years*, Coleraine: Centre for the Study of Conflict, University of Ulster.

Sterne, J. (2012), 'Introduction: Sonic Imaginations', in J. Sterne (ed.), *The Sound Studies Reader*, 1–18, Abingdon: Routledge.

Strauss, A., and J. Corbin (1990), *Basics of Qualitative Research: Grounded Theory Procedures and Techniques*, London: Sage.

Tashakkori, A., and C. Teddlie (1998), *Mixed Methodology: Combining Qualitative and Quantitative Approaches*, vol. 46 of *Applied Social Research Methods*, Thousand Oaks, CA: Sage.

Taussig, M. (1993), *Mimesis and Alterity: A Particular History of the Senses*, London: Routledge.

Taylor, J. (2010), 'Introduction', *Ubu and the Truth Commission*, from the production by William Kentridge and the Handspring Puppet, Cape Town: University of Cape Town Press.

Tedlock, B. (1991), 'From Participant Observation to the Observation of Participation: The Emergence of Narrative Ethnography', *Journal of Anthropological Research*, 47 (1): 69–94.

Thich N. H. (2015), *Silence: The Power of Quiet in a World Full of Noise*, London: Rider.

Thompson, E. (2001), 'Empathy and Consciousness', *Journal of Consciousness Studies*, 8 (5–7): 1–32.

Tilly, C. (2004), *Social Movements, 1768–2004*. Boulder: Paradigm.

Tripp, C. (2013), 'Performing the Public: Theatres of Power in the Middle East', *Constellations*, 20 (2): 203–16.

Turino, T. (2008), *Music as Social Life: The Politics of Participation*, Chicago, IL: University of Chicago Press.

UNESCO (2017), 'The Importance of Sound in Today's World: Promoting Best Practices', General Conference 2017, Paris, pub. 39 C/49, 1–4. Available online: https://unesdoc.unesco.org/ark:/48223/pf0000259172.locale=fr (accessed 4 January 2022).

Urbain, O. (2008a), 'Introduction', in O. Urbain (ed.), *Music and Conflict Transformation: Harmonies and Dissonances in Geopolitics*, 1–9, London: I.B. Tauris.

Urbain, O. ed. (2008b), *Music and Conflict Transformation: Harmonies and Dissonance in Geopolitics*, London: I.B. Tauris.

Urban, E. (2011), *Community Politics and the Peace Process in Contemporary Northern Irish Drama*, Oxford: Peter Lang.

Urban, E. (2020), 'Fractured Liminality in Kabosh's Green and Blue and Lives in Translation', in A. Goarzin and M. Parsons (eds), *New Cartographies, Nomadic Methodologies: Contemporary Arts, Culture and Politics in Ireland*, 127–44, Oxford: Peter Lang.

van der Vaart, G., B. van Hoven and P. P. Huigen (2018), 'Creative and Arts-Based Research Methods in Academic Research: Lessons from a Participatory Research Project in the Netherlands', *Forum: Qualitative Social Research*, 19 (2): 1–30.

Varoutsos, G. 'Peace Wall Belfast: Spatial Audio Representation of Divided Spaces and Soundwalks'. doi: https://doi.org/10.5281/zenodo.3898717 (accessed 17 June 2020).

Voegelin, S. (2010), *Listening to Noise and Silence: Towards a Philosophy of Sound Art*, New York: Continuum.
Voegelin, S. (2011), 'Pulse', in H. Schulze (ed.), *The Bloomsbury Handbook of the Anthropology of Sound*, 107–10, New York: Bloomsbury Academic.
Voegelin, S. (2021), 'Sonic Methodologies of Sound', in M. Bull and M. Cobussen (eds), *The Bloomsbury Handbook of Sonic Methodologies*, 269–80, London: Bloomsbury Handbooks.
Wang, Q., S. Coevens, R. Siegesmund, and K. Hannes (2017), 'Arts Based Methods in Socially-Engaged Research Practice: A Classification Framework', *Art/Research International: A Transdisciplinary Journal*, 2 (2): 5–39.
Wardani, F. (2020), *Interview with Julie Norman*, Beirut: Lebanon.
Wlodarski, A. L. (2010), 'The Testimonial Aesthetics of *Different Trains*', *Journal of the American Musicological Society*, 63 (1): 99–141.
Yamagishi, T. (2001), 'Trust as a Form of Social Intelligence', in K. S. Cook (ed.), *Trust in Society*, 121–47, New York: Russel Sage Foundation.
Yeats, W. B. (1955), *Autobiographies*, London: Macmillan.
Yee, S. (2018), *So I Can Breathe This Air* (unpublished script; based on anonymous interviews with members from The Rainbow Project's Gay Ethnic Group).
Yin, R. K. (2003a), *Applications of Case Study Research*, 2nd edn, vol. 34 of *Applied Social Research Methods Series*, Thousand Oaks, CA: Sage.
Yin, R. K. (2003b), *Case Study Research: Design and Methods*, vol. 5 of *Applied Social Research Methods Series*, Thousand Oaks, CA: Sage.
Zembylas, M. (2018), 'Reinventing Critical Pedagogy as Decolonizing Pedagogy: The Education of Empathy', *Review of Education, Pedagogy, and Cultural Studies*, Special Issue, 'Pedagogy of the Oppressed: 50 Years', 40 (5): 404–21.
Zigon, J. (2009), 'Within a Range of Possibilities: Morality and Ethics in Social Life', *Ethnos*, 74 (2): 251–76.

Videography

Buffalo String Works (2021b), 'How Was "They Would Only Walk" Composed?' YouTube video, 01:45. 18 February 2021. Available online: https://www.youtube.com/watch?v=YiLeL6YDFYQ (accessed 19 July 2021).
Tinderbox Theatre Company (2019), 'Clips from Ubu the King performance', 19 February 2019. Available online: https://www.qub.ac.uk/research-centres/SoundingConflict/ResearchFindings/ReportsandPublications/ProjectMonograph2022.html (accessed 16 September 2022).
Water, Earth & Sky (2021), dir. Sunny Liu. New York: Arium TV.

Contributors

Fiona Magowan

Fiona Magowan is Professor of Anthropology and Fellow of the Senator George J. Mitchell Institute for Global Peace Security and Justice at Queen's University, Belfast. As Principal Investigator of the Partnership for Conflict Crime and Security Research (PaCCS)–funded project 'Sounding Conflict: From Resistance to Reconciliation' (2017–22), hosted by the Mitchell Institute, Fiona has been leading a team of seven Queen's staff and ten partner organizations researching sound, music and storytelling across the Middle East, Brazil and Northern Ireland. She has conducted long-term research in north-east Arnhem Land, Australia, Queensland, and South Australia and recent projects in Brazil and Mozambique. Her ethnographic research focuses on three interconnected areas: music, sound and movement; art, emotion and the senses; and religion, identity and social transformation. She has published seven books including, *Christianity, Conflict, and Renewal in Australia and the Pacific* (Brill 2016, co-edited with Carolyn Schwarz); *Performing Gender, Place, and Emotion* (Rochester 2013, co-edited with Louise Wrazen); and *Melodies of Mourning: Music and Emotion in Northern Australia* (Oxford, James Currey 2007).

Julie M. Norman

Julie M. Norman is an Associate Professor of Politics and International Relations at University College London (UCL). Her research lies at the intersection of human rights, security and resistance in protracted conflicts, with a focus on the Middle East and North Africa. She is the author of *The Palestinian Prisoners Movement: Disobedience and Resistance* (Routledge 2021) and three books on nonviolent resistance, including *Understanding Nonviolence* (Polity 2015; with Maia Carter Hallward) and *The Second Palestinian Intifada: Civil Resistance* (Routledge 2010). She has also published widely on conflict and development, political detention and violent extremism, and she is a frequent political analyst on BBC, CNN, Al Jazeera and other media outlets. Prior to joining UCL, she was a Research Fellow at the Senator George J. Mitchell Institute for Global Peace, Security and Justice at Queen's University Belfast.

Ariana Phillips-Hutton

Ariana Phillips-Hutton joined the Senator George J. Mitchell Institute for Global Peace, Security and Justice at Queen's University Belfast in 2021 as a Research Fellow on the *Sounding Conflict: From Resistance to Reconciliation* project. Her research centres on the philosophy, performance and politics of contemporary music, with particular interests in violence, conflict transformation and musical ethics. Recent publications include articles in *Twentieth-Century Music*, *Popular Music*, the *Journal of the Royal Musical Association* and the *Journal of the British Academy*. She is also the author of *Music Transforming Conflict* (Cambridge University Press, 2020) and associate editor for the *Oxford Handbook of Western Music and Philosophy* (Oxford University Press, 2020). She is is Lecturer in Global Critical and Cultural Study of Music at the University of Leeds.

Stefanie Lehner

Stefanie Lehner is Senior Lecturer in Irish Literature and Fellow of the Senator George J. Mitchell Institute for Global Peace, Security and Justice at Queen's University, Belfast. Her current research explores the role of the arts, specifically performance, in conflict transformation processes, with a focus on the Northern Irish context. She also researches and teaches on representations of trauma and memory in (Northern) Irish drama, fiction, film and photography. She is author of *Subaltern Ethics in Contemporary Scottish and Irish Literature* (2011), and her work has been published in *Contemporary Theatre Review*, *Irish Review*, *Irish Studies Review*, *Irish University Review* and *Nordic Irish Studies*, as well as collections published, among others, by Cambridge and Oxford University Presses. She has worked on the Partnership for Conflict Crime and Security Research/Arts and Humanities Research Council (PaCCS/AHRC)–funded 'LGBTQ Visions of Peace' project as well as the 'Sounding Conflict: From Resistance to Reconciliation' project.

Pedro Rebelo

Pedro is Professor of Sonic Arts at Queen's University, Belfast, composer, sound artist and performer. Pedro has led participatory and socially engaged art projects involving communities in Belfast; favelas in Maré, Rio de Janeiro; travelling communities in Portugal; and a slum town in Mozambique. This work has resulted in sound art exhibitions at venues such as the Metropolitan Arts Centre, Belfast; Centro Cultural Português, Maputo; Espaço Ecco in Brasilia; Parque Lage and Museu da Maré in Rio; Museu Nacional Grão Vasco; Golden Thread Gallery; Whitworth Gallery Manchester; Convento de São Francisco Coimbra; and MAC Nitéroi. His music has been presented in venues such as the Melbourne Recital Hall, National Concert Hall Dublin, Queen Elizabeth Hall, Ars Electronica, Casa da Música, and in events such as Weimarer Frühjahrstage fur zeitgenössische Musik, Wien Modern Festival, Cynetart and Música Viva. His work as a pianist and improvisor has been released by Creative Source Recordings, and he has collaborated with musicians such as Chris Brown, Mark Applebaum, Carlos Zingaro, Evan Parker and Pauline Oliveros, as well as artists such as Suzanne Lacy. Pedro has been appointed Director of the Sonic Arts Research Centre in 2021.

Partners

Buffalo String Works

Buffalo String Works' mission is to ignite personal and community leadership through accessible, youth-centred music education. We provide rigorous music instruction and a creative home for refugee, immigrant and historically marginalized youth living in Buffalo, NY. We recognize the significance of music as a universal language, and by lifting up the voices of our students and parents, we cultivate youth to be agents of social change.

Our vision: Neighbourhoods united by empowered young musicians.

For more information, please visit us online at buffalostringworks.org

Follow us on Facebook and Instagram @buffalostringworks

Fighters for Peace

Fighters for Peace (FFP) is a Lebanon-based non-governmental organization founded in 2014 by ex-combatants from the Lebanese civil war with the aim to foster a national reconciliation process and prevent violent extremism in the Middle East. The organization provides a safe space in which ex-fighters and former extremists can seek support for their personal transformation processes. FFP members are living examples that even the fiercest partisans can transition from war to peace. For those who have engaged in war or violent extremism, there is a dire need to gain new perspectives for their lives in new, non-violent leadership roles. FFP members meet those who are currently engaged in violence, helping them walk away from conflict or transition from past to new roles. FFP supports them in finding an alternative sense of purpose and belonging, trains them and then puts them in the front line for the fight for peace and the prevention of violent extremism. They reach out to fellow combatants, youth and the general public.

http://fightersforpeace.org/

Laban

Laban is a non-profit civil organization founded in Beirut in 2009 based on improvisational theatre. Laban utilizes Improvisation Theatre, Interactive Theatre, Playback Theatre, and Theatre of the Oppressed in its components of civil work to engage with social justice issues in the Lebanese society. Laban works on two tracks: first, spreading improvisational theatre in Lebanon through performances, training and creating a community of improvisational theatre artists, and second, using improvisational theatre for civil society activism through training, raising awareness through performances, interactive performances and forms and tools especially invented for specific campaigns and projects with theatre as the main component.

Follow: @LabanLive

Mary Kouyoumdjian

Mary Kouyoumdjian is a composer and documentarian. As a first-generation Armenian American and having come from a family directly affected by the Lebanese Civil War and Armenian Genocide, she draws on her heritage, documentary and background in experimental composition to blend the old with the new. A strong believer in freedom of speech, her work often integrates recorded testimonies with resilient individuals and field recordings of place to invite empathy around social and political conflict. Kouyoumdjian is on the composition faculty at the Boston Conservatory at Berklee, is a cofounder of the annual conference New Music Gathering and is published by Schott's PSNY.

Musicians Without Borders

Musicians Without Borders is a leader in using music for peacebuilding and social change. For over twenty years, Musicians Without Borders has worked with people and communities affected by war, conflict and displacement around the world. Projects range from rock music schools to children's orchestras, from music therapy to hip hop, and from songwriting to samba. Musicians Without Borders has established long-term, collaborative programmes in the Balkans, the Middle East, Central East Africa, Europe and Central America.

Sharing lessons from the global grassroots programmes, Musicians Without Borders trains musicians and activists to create change in their communities through musical leadership grounded in principles of nonviolence and universal human rights.

www.musicianswithoutborders.org

Kabosh

Founded in 1994, Belfast-based theatre company Kabosh gives voice to sites, space and people through reinventing and reinterpreting the ways in which stories are told. Kabosh is committed to challenging the notion of what theatre is, where it takes place and who it is for. A large percentage of the company work deals with the legacy of conflict and is toured internationally. Paula McFetridge has been Artistic Director since 2006. She is a fellow of Salzburg Global Seminar Session 532 'Peacebuilding through the Arts' and was made Belfast Ambassador in recognition of utilizing the arts to tackle difficult issues. For further information, see www.kabosh.net

TheatreofplucK

Northern Ireland's first publicly funded gay theatre company aims to produce quality theatre for everyone, but with a queer slant. They especially explore issues of LGBT+ identities in Ireland. Pluck's first incarnation was a queer theatre/dance hybrid company in Philadelphia formed by director/designer Niall Rea and performer Karl Schappell. The name Pluck was chosen to honour the memory of Christopher Hawks who had lived with Rea and Schappell but who had died from AIDS shortly before the company was formed. Hawks would have undoubtedly been a core company member, and his catch phrase of "Oh Pluck!" when someone did or said something special or out of the ordinary or brave seemed to embody the ethos of our attitude to performance. When Rea returned to his native Belfast, they began to work together on *Automatic Bastard* using a space access grant from the Lyric Theatre. Following rave reviews and now called TheatreofplucK, they received their first Arts Council of Northern Ireland grant and soon after were asked to join the Hatch artist residence programme at Belfast's new flagship arts venue the Metropolitan Arts Centre, making four new queer theatre works during the residency. They went on to open their own queer performance venue The Barracks in Belfast's cultural Cathedral Quarter and have toured productions to London, Philadelphia, South Carolina, Amsterdam and Dublin.

http://www.theatreofpluck.com

Tinderbox Theatre Company

Tinderbox Theatre Company is a leading arts organization specializing in contemporary theatre practice in Belfast, Northern Ireland. Since 1988, Tinderbox has nurtured and championed new writing, producing performances from Northern Irish writers and theatre makers to international acclaim. Since 2016, under new artistic direction by Patrick J O'Reilly, Tinderbox introduced a new vision of artistic practice and engagement to focus on a creative process that liberates the imagination for artists and participants, challenges the social barriers of access and engagement and most importantly champions authenticity for individuals and communities. We are inspired by creative innovation, experimentation and contemporary theatre. In 2021, with the new appointment of Meg Magill as Producer, Tinderbox developed a three-year strategic programme called Playground which celebrates the artist, local spaces and the environment in the effort to create sustainable eco-theatre practice through our Off the Grid initiative.

https://www.tinderbox.org.uk

Museu da Maré

Founded in 2006, the Maré Museum emerged out of a desire to preserve memories and histories of Maré, Rio de Janeiro, and resulted in the integration of different social agents, guaranteeing its plurality.

The museum's work aims to overcome stigmas in relation to favelas as well as broaden the role of the museum in contemporary reality. The museum is not a place to store objects or create a cult of the past. It is a place of life, conflict and dialogue immersed in the past while looking to the future in relation to current communities, their conditions and identities and cultural and territorial diversity.

Currently, projects involve around 500 people with the state schools in the area being the main partners. The long-term permanent exhibition 'Times of Maré' has received over seventy thousand visitors. Maré Museum has been awarded the most prestigious prizes in this area in Brazil.

https://www.facebook.com/museudamare

Index

Aboultaif, E.W. 62
Abu Dhabi 167
acousmatic music 32
acoustemology 3
acoustic ecology 30
Acoustic Justice (LaBelle) 35
acoustic politics of space 89–91
Acoustic Territories (LaBelle) 79
ACT for the Disappeared 56
activism 13, 19, 37, 39, 48, 51, 53–4, 57, 181, 185
affective resonance 75, 81
   definition 72
   listening 73
   resilience 73
Afghanistan 22, 100–1
Afghanistan National Institute of Music 99–100, 118
Aghavni 97
Alaimo, E. 104
Alarm Will Sound 102–3
America 109
anamnesis effect 83, 125, 134, 159–60
Annette 78–80, 84–6, 89–90
anthropology 2
Arendt, H. 130
Armenian Genocide (1915–23) 97
arpilleras 12
artistic individuation 12
artistic research 42
Asbury Hall, Buffalo 107
Assaad, N. 68
Audio-Vision: Sound on Screen (Chion) 32
audiovisual litany 5
audio walk 145–6, 175
Augoyard, J.-F. 125, 159–60

Baaz, M. 48, 52
Bagnall, C. 140
Baker, G. 26
Bangkok 106

Barenboim, D. 126
Barron, V. 104
Bartels, D. 177
Barthes, R. 4
Battle of the Bogside 75
Beirut 19, 38, 56–8, 66, 98
Belfast 24, 44, 140, 152–3
Belfast EastSide Arts Festival, 2018 144
belliphonic sounds 124–5, 129
Benjamin, W. 140
Berninger, M. 107
Bethlehem 20, 93
Beuys, J. 167
Bishop, C. 8–9, 154
Bloody Sunday 16, 75
Bloomfield, D. 15
Bogart, A. 7
Bombs of Beirut 97–8, 109
Born, G. 34
Bosnia 85
Botafogo 161
Bourriaud, N. 4, 7, 9
Brazil 11, 25, 29, 44, 69, 158–9, 163, 183
Brody, M.J. 100
Brown, R. 125
Browne, B. 55
Bryars, G. 109
BSW, *see* Buffalo String Works
Buddha 165
Buffalo 22, 99, 104–8
Buffalo Chamber Players 22, 107
Buffalo String Works (BSW) 22, 98, 103–8, 112, 117, 180
Bull, M. 26
Burma 99, 106, 118
Butler, J. 54

Cage, J. 30, 34, 36
California 97
Caracelli, V. 36
carioca funk 25

Carnegie Hall, New York City 99
causal mode, listening 32
Cavarero, A. 44, 97, 110–11
Charlie Brown 157
Children of Conflict 97
Chile 53
Chion, M. 32
Clandinin, J. 39
Clifford, J. 3
Clinton, B. 16–17
(co-)creative methods 40–2
Cohen, C. 15, 125
Coldplay 157, 162
collaboration 7, 97–8, 102, 151
    Laban theatre group 57, 63
    MWB 20
    performative 10
    Sounding Conflict project 162–72
collective memory 56, 65
Columbia 102
Columbia University 108
Come Out 108–9
Committee of the Families of the Disappeared 56
compassionate listening 34–5
conflict; *see also Sounding Conflict*
    agitatory noises of 124–5
    intergenerational memory of 65
    voice, role of 49
conflict transformation and peacebuilding
    artistic research 42
    case studies 37
    challenges and limitations 42–4
    (co-)creative methods 40–2
    listening 30–6
    mixed methods 36–7
    narrative interviews 39–40
    participant observation 38–9
    practice-based research 42
connectedness 21, 177, 179
Connelly, M. 39
Connor, S. 3, 33, 73
covert–overt spectrum, resistance 53, 60, 69
covert resistance 49
Covid-19 pandemic 40, 43, 106, 117
creative arts, sound-based 5
creative strategies, creative arts 184–6

creative tension 79–81, 89
creativity 21, 41–2, 86–7, 89, 93, 95, 112, 173
    MWB's approach 20, 72–4
    productive 72, 93
    resilience 72–4, 79–82, 86–94
    socially engaged arts interventions 151–72
    strategic 12
Csikszentmihályi, M. 80
Cultúrlann Uí Chanáin 20, 75
Czechoslovakia 12

Daughtry, J.M. 124–5
Denmark 117
Derrida, J. 111
Derry/Londonderry 44, 75, 77, 79, 85–6
Derry/Londonderry cultural organization 20
Dessner, A. 107
d'Estree, T.P. 49, 52
dialogue 178
    Playback theatre 64
    sessions, FFP 56
Different Trains 109
Dillon, S. 15
dissonant harmony 26, 125–7
Dolan, J. 148
Donaghey, J. 76
Dorsey, G.A. 107
Duchamp, M. 33
Duncombe, S. 54
Dun Laoghaire County Council 90

Eco Week 2017 programme 90
Egoyan, E. 97
Eidsheim, N.S. 36, 115
Einwohner, R. 12, 48–51
Electrical Walks (Kubisch) 33
Emery, D.B. 12
emotion 7, 25, 40–2, 58, 60, 63, 65, 67, 80, 82, 84, 89, 93–4, 108, 117, 122, 127–8, 132, 135, 141–2, 148, 160, 174, 179
emotional stimulus 122
empathy
    intercultural 92–4
    narration 179–81
    Playback 67

relationality 14
witnessing 6
Erikson, E. 74
Eritrea 106
Erlmann, V. 163
ethical awareness 89–91
ethnographic composition 98, 107, 112, 114, 181
Europe 109, 210
everyday resistance
  de facto gains 50
  definition 50
  and power 51
  repertoires of contention 51
  spatial and temporal implications, resistance 51–2
  temporality 51
extramusical sounds 4

Feld, S. 2
FFP, *see* Fighters for Peace
Fiddian-Qasmiyeh, E. 31
Fighters For Peace (FFP) 18–20, 38, 55–7, 61, 63, 182
First Intifada 13
Fischer-Lichte, E. 123, 127–8, 136, 142–3
Fluxus movement 167
Foerch, C. 58–9
forgiveness 15, 23, 184
For More than One Voice (Cavarero) 111
Forum Theatre 19
Foster, V. 45
Foucault, M. 48–9
freedom
  creative 95
  participatory music making 78
Freud, S. 6
Funk Carioca 159

Gallagher, M. 26
Ganz, M. 52
García, Y. 22, 98, 106, 116–18
Garda Síochána 23, 121, 123
Gay Ethnic Group (GEG) 24, 123, 144–5
GEG, *see* Gay Ethnic Group
Gershon, W. 153
Giddens, A. 74
Good Friday Agreement 122, 134, 145, 185

Gopinath, S. 109
Graham, W. 36
Green and Blue 23, 122, 124–37, 174
  affiliative reconciliation 133
  anamnesis 134
  belliphonic sound 132
  interfering radio noises 132
  liminality, sense/state of 136
  perceptions 135–6
  sounds of friendship 132–4
  transformative outside/inside perceptions 134–7
Greene, J. 36
Grønseth, A.S. 87

Haile, S. 181
Hands across the Divide statue 16
Hanh, T.N. 34–5
Hannigan, L. 107
Hariri, S. 57
harmony-as-consonance 126
Harron, M. 16
Havel, V. 12
Healing Through Remembering 23, 123
hearing 3
Heidemann, B. 122
Hezbollah 57–8
hip hop 25, 167
Hollander, J. 12, 48–51
Hondros, C. 97
Hong Kong 13
Howell, G. 26, 126

I Can Barely Look 97
Ihde, D. 31–2
improvisation 76–9, 87
individual-collective spectrum, resistance 52–3
intercultural empathy 92–4
International Committee of the Red Cross (ICRC) 56
interpersonal resilience
  sonic relationality 82
  transformative engagement 81
interventions 145, 147; *see also* socially engaged arts
  arts-based 68
  performative 145

Playback 56–7, 62–4, 67–8
  psychological resistance 57
  and resistance 59
interviews
  narrative 39–40
  open-ended 108
  semi-structured 76
intrapersonal healing 63–4
intrusive noises 23, 122, 127, 129–30, 133, 141, 149, 174
Iran 13
Ireland 24, 90, 211
Irish Republican Army (IRA) 121
Israel 57
It's Gonna Rain 108–9

James, R. 5
Jammal, G. 63–7
Jarry, A. 24, 123, 138–40
Johansson, A. 51
Johnson, M. 165
Jordan 167
Jorgensen, D.L. 38
Josephides, L. 87

Kabosh 23, 121, 123–37, 148–9, 174
Kabul 22, 99
Kalashnikov fire 125
Kee, H. 106
Keelan, E. 55
Kendrick, L. 124
Kennedy Center, Washington DC 99
Khader, S.J. 115
Kim-Cohen, S. 33
Klein, J. 42
Kneecap, 2019 167–8
Korum, S. 26, 126
Kouyoumdjian, M. 22, 44, 97–110, 112–17, 180–1
Krog, A. 16
Kronos Quartet 98
Kubisch, C. 33–4

Laban Lactic Culture 55
Laban performance art theatre 19–20, 38, 183
LaBelle, B. 35, 79, 82, 85, 89
Lacy, S. 9

Lakoff, G. 165
Lanz, B. 107
Larkin, C. 62
Laurence, F. 180
Lebanese civil war (1975–89) 55, 62, 97–8
Lebanon 18, 21, 38, 185
  resilience and reconciliation 61–2
  sound design 167
Lebanon case study
  activism 57
  FFP 55–7, 61
  Laban 56–7
  music 58
  narrative-driven outreach activities 56
  Playback theatre programme 56–7
  resistance 57
  storytelling 56, 58–60
  theatre 60–1
  2019 uprising 57
Lederach, J.P. 91, 174
Lee, D. 148
Lefebvre, H. 166
Lehner, S. 123
Leonard, J. 122
Lewis, R. 137
Lilja, M. 48, 52
limbic resonance 2
liminality 123, 128, 140, 143, 145
listening 10, 14, 26
  and affective resonance 73
  causal mode 32
  cognitive engagement 36
  compassionate 34
  deep 34–5
  ethics of 173–6
  guided 162
  vs. hearing 33–4
  immersive 144–8
  intentional action *vs.* passive causality 33
  as methodology and method 30–1
  mindful 34
  modes 32
  non-cochlear sonic art 33
  perception and experience 31–6
  performative 35–6
  reduced 32
  semantic 32
  social 162

socially relational nature 35
sound and space 32–3
Listening and Voice (Ihde) 31
Listening (Nancy) 73
LISTEN (Neuhaus) 33
London 98

Magowan, F. 76, 156
Maister, N. 102
Malaysia 106
Mannes School of Music 100
MAP Fund 102
Maré favela complex 152, 183
Maxwell, J. 37–8
McFetridge, P. 23, 123, 128, 131
McIvor, C. 128
McKeown, L. 23, 123
McMurray, P. 17
MC Orelha 2014 167
McRae, C. 35–6
Meireles, M. 24, 155, 163
methodologies, conflict transformation and peacebuilding
　artistic research 42
　case studies 37
　challenges and limitations 42–4
　(co-)creative methods 40–2
　listening 30–6
　mixed methods 36–7
　narrative interviews 39–40
　participant observation 38–9
　practice-based research 42
Metropolitan Arts Centre, Belfast 140
Mexico 22, 106
Middle East 29, 163
Middleton, I. 74
Missouri 102
Mizzou International Composers Festival 102
Modern Auditory I (Connor) 73
moral imagination 91, 174, 180
Morocco 106
Morro do Timbau 158
Mourad, K. 102
mourning 6–7
Moyra 80
Multiple Journeys (of Belonging) 24, 123
muqāwamah 57

muro da vergonha 152
Museu da Maré 155–6, 161
music
　education to refugees 22
　facilitation 76
　and harmony 126
　Lebanon case study 58
　making 2, 4, 11, 15, 21, 72–3, 76, 78, 84, 86, 88, 92, 94–5, 116–17, 162
　peacebuilding 14
　relational aesthetics 114–16
　and social transformation 115
musical education, refugees 22
Music Bridge programme 75–7
musicians 10–11, 43, 51–2; *see also* Musicians Without Borders
　from Buffalo 107
　Kouyoumdjian, M. 22, 44, 97–110, 112–17, 180–1
Musicians Without Borders (MWB)
　collaborative participatory action approach 176
　community healing 184
　creative response/proactivity 80
　creative tension 79–81
　empathic affinity 79
　empathic empowerment 73
　ethical awareness 89–91
　graphic elicitation method 41
　improvisation, structuring of 76–9
　inclusivity 84–6
　intercultural empathy 92–4
　interpersonal resilience 81–2
　listening 73
　mimetic excess 77–8
　Music Bridge programme 75–6
　nonviolence 20
　participant observation 38
　positive reinforcement 82–4
　resilience 184
　safety and inclusivity 77
　sound, role of 72
　training programmes 71
　transformational resilience 86–9
　trust, improvisatory practice 74–5
musicking
　empathic empowerment 72
　as ethical reasoning and care 176–9

improvisatory 75
nonverbal 79
participatory 2, 21, 72, 92, 94, 176
resilience 13–15
safe space for 87–8
MWB, *see* Musicians Without Borders
Myers, M. 146–7

Nancy, J.-L. 35, 73
narrative
  empathy 179–81
  interviews 39–40
  quilt squares 12
negative liminality 123
Neuhaus, M. 33–4
New England Conservatory, Boston 99
New York City 22, 57, 99–101, 106, 117
Nielsen, T.R. 144
Nitsch, H. 167
Noddings, N. 178
noise
  agitatory 124–5
  intrusive 124–5, 127
  somatic impact 124
non-musical sounds 4
nonverbal music exercises 73
nonverbal safe spaces 76–9
nonviolence 20, 48, 56, 59, 209, 211
Norman, J. 54
Northern Ireland 11, 18, 20–1, 23–4, 29, 69, 75–6, 93–4, 122–3, 138, 145, 153, 163, 167, 184–5
Northern Irish conflict 123, 131
Norway 87
Norwegian school music education programme 14
Nova Holanda 156
Novak, D. 3, 5

Oliveros, P. 34–5
open-ended interviews 108
oral recordings 109
O'Reilly, P.J. 24, 123, 163–4
othering 64
overt resistance 49
Oxford English Dictionary (OED) 15

Palestine 20–1, 93–4, 184

Paper Pianos: 'You Are not a Kid' 22, 98–103, 107–8, 112–13, 115–17, 119
Paris 139
participation
participant observation 38–9
participatory art 8
participatory musicking 2, 21, 72, 92, 94, 176
participatory song writing 21
peacebuilding 11, 14; *see also* conflict transformation and peacebuilding
  music and 126
  MWB 20
PEACE IV Programme 23
Peace Wall, Belfast 152
perceptual multistability 142
performative listening 35–6
Perry, R.R. 107
Phelan, M. 138
Phenomenologies of Sound (Ihde) 31
Philadelphia 24
Phillips-Hutton, A. 180
phonomnesis 83–4
Pinochet 13, 53
Playback Theatre 19, 56–7, 62–4
  empathy 67
  and war memory 67
Playboy of the Western World, The 139–41
Post-Agreement theatre 122
power to truth *vs.* truth to power 181–4
practice-based research 42
praxis, modalities of 177
Prior, J. 26
protest music of the 1960s 12

Rabinger, M. 112
Rae, N. 24, 148
Ramallah 93
Rancière, J. 175
rap 25, 167
Rathcoole 134
Rebelo, P. 24, 155, 163
reconciliation 6, 67–8
  definitions 15–16
  dissonant sounds of 23–4
  handshake of 16
  performative dimensions 15–17

performed sound, transformative power
   of 127–9
  and transitional justice 68
reduced listening 32
Reich, S. 108–9, 112
Reilly, S. 107
Relating Narratives (Cavarero) 97
relational aesthetics 114–16
Relational Aesthetics (Bourriaud) 7
relational art 7
relational/interactional nature,
   resistance 48–50
remediation 113–14
  cultural studies 17
  refugee stories 21–3
  sonic 17–18
Rennie, T. 155
resilience
  creative practice 20–1, 71–94
  creative tension and productivity 79–81
  definition 14
  ethical awareness 89–91
  inclusivity 84–6
  intercultural empathy 92–4
  interpersonal 81–2
  intrapersonal 94
  misrecognition 13
  Music Bridge programme 75–6
  MWB 20–1, 72–4
  nonverbal safe spaces 76–9
  positive reinforcement, sonic
    effects 82–4
  resilient community 71
  structuring improvisation 76–9
  through musicking 13–15, 72, 74–5,
    79–84, 87–9
  transformational 86–9
  trust 74–5
resistance
  artistic individuation 12
  arts-based 37, 54
  community and society 65–8
  covert–overt 53
  and creative practice 12–13
  definition 12
  everyday 50–2
  FFP 18–20, 55–7, 61, 63
  individual–collective spectrum 52–3

  interpersonal 64–5
  intrapersonal healing 63–4
  Laban 19–20
  Lebanon case study 55–61
  performing frontline 18–20
  post–civil war Lebanon and forceful
    amnesia 62–3
  power, role of 48–9
  process (expression)–outcome
    (impact) 53–4
  relational/interactional nature 48–50
  resilience and reconciliation 54–5, 61–2
  sonic practices 13
  sound-based 51–2
  spectral elements 52–4
  voice/sound 49–50
  and vulnerability 54
Resnick, Y.N. 99, 103, 117
Reunion statue 16
Richards, S. 138–9
Ricoeur, P. 6–7, 45
Rio de Janeiro 25, 152, 155, 167
Robbins, J. 87
Roesner, D. 124
Rollefson, J.G. 167
Roman, J. 107
Rost, K. 122, 124, 127, 133, 141, 143
Royal Ulster Constabulary (RUC) 121, 123,
   130, 132, 134–5
Rwanda 85

Sakakeeny, M. 3, 5
Sarmast, A. 118
Saville Inquiry, the 75
Schaap, A. 130
Schaeffer, P. 32
Schafer, R.M. 4, 10, 30
Schulz, M. 48, 52
Scott, J. 50, 53
sectarianism 75
semantic listening 32
semi-structured interviews 76
*Sensing Sound: Singing and Listening as
   Vibrational Practice* (Eidsheim) 36
Sharp, G. 48
Silent Cranes 97
Sinn Féin 121, 129
Small, C. 72

social capital 176
social cohesion 59, 182
social healing 59
social intelligence 75
socially engaged arts
 criticism 9
 participatory art 8
 relational art 7
 Som da Maré project 155–62
 Sounding Conflict project 162–72
So I Can Breathe This Air 24, 123, 144–8, 175
 attack, live actor 147
 audio walk, performative 145–6
 LGBTQ scene, Belfast 148
 liminality 145
 shared responsibility 144
 sociopolitical injustice 147–8
 verbatim 145
Som da Maré project 25
 activities 155
 anamnesis 159–60
 anticipation 160
 art interventions 155
 childhood memory 157
 community and sound 156
 guided listening 162
 listening, active and political act 159
 Museu da Maré 155–6
 reflective time 157
 sound recording 158
Somos CV 167
Sonic Agency 35
sonic ethnography 26, 153
sonic practices
 effects, positive reinforcement in 82–4
 listening 31
 non-cochlear 33
 of resistance 13
 socially engaged 13
 Som da Maré project 155–62
sound; *see also* voice
 agitatory noises, conflict 124–5
 belliphonic 124–5, 129
 creative arts 5
 effects, theatre performances 122
 extramusical 4
 of friendship 132–4

hearing 3
indivisible cosmos 10
intergenerational relationships 116
intrusive 124–5
limbic resonance 2
 as material and metaphor 3–5
 memory of 125
 multi-layered sonic character 4
 non-musical 4
 participatory arts 4–5
 reconciliation 15–17, 125–7
 of reconciliations 125–7
 relationship to memory 83
 remediation 17–18
 resilience 13–15, 20–1
 resistance 12–13, 18–20
 sense of nostalgia 157
 socially engaged arts and sonic sphere 7–9
 sources of 109
 and space relationship 32–3
 spaces of deafness 5–6
 and storytelling 6–7
 thinking 2–3
 transformative power 127–9
 vibrational reality of 35
 voice, levels of 49–50
 word and interpretation 9–11
sound art 2, 18, 29, 33, 152, 186; *see also Sounding Conflict*
sound design 25, 122, 128, 153, 164, 167
Sounding Conflict: A Performance in Five Acts 24–5
 act I: resistance 168
 act II: reconciliation 169
 act III: resilience 169–70
 act IV: resilience 170–1
 act V: resistance 170–1
 act VI: coda 171–2
 creative and production blueprint 156
 filmed performance 163
 hip-hop practices 167
 sound design 167
soundscapes 10, 27
 for change 186–7
 definition 186
 social and political change 187
 sonic realities 186

South Africa 16
South Dublin County Council 90
South Korea 106
South Sudan 13
spaces of deafness 5–6
spectral elements, resistance
   covert–overt 53
   individual–collective 52–3
   process (expression)–outcome
      (impact) 53–4
speech-melody 109
Sprenglemeyer, L. 107
state-enforced amnesia 57, 62–3, 65,
   68, 182
Sterne, J. 5
Stormont Castle, Belfast 16
storytelling 6–7, 19, 29
   and compositional process 111–12
   Different Trains 109
   documentary music video 109
   effects 116–19
   ethnography, ethics and
      composition 107–12
   human connections 110
   humanity 58
   Lebanon case study 56, 58–60
   musical 111–12
   open-ended interviews 108
   Paper Pianos: 'You Are not a Kid' 98–103,
      107–8, 112–13, 115–17, 119
   reconciliation 115
   relational aesthetics 114–16
   relational asymmetries 110–11
   remediation 113–14
   social cohesion 59
   They Would Only Walk 98, 103–
      7, 115–19
strategic creativity 12
2 Suitcases 97
Synaptic 2018, The 167
Synge, J.M. 139–40
Syria 66
Syrian refugee crisis 97

Tabar, P. 62
Taif Agreement, the 62
Taliban 100–1
Taussig, M. 78, 86

Taylor, J. 139
thawra 57
theatre
   intrusive sounds 124–5, 127
   Lebanon case study 60–1
*Theatre Noise: The Sound of
   Performance* 124
TheatreofplucK 23–4, 123, 144–8,
   175
Theatre Sound 122
They Will Take My Island 97
They Would Only Walk 22, 98, 103–7,
   115–19
This Should Feel Like Home 97
Those You Pass on the Street 23, 121–37,
   174
   aural theatrical space 130–1
   dissonant voices 129–32
   doorbell, noise of 127–8
   feedback forms, audience 131–2
   reconciliation 130–1
Tilly, C. 51
Time for Three 107
Tinderbox 23–4, 123, 137–44, 163,
   175
Torgue, H. 125, 159–60
transformational resilience
   ethical reflexivity 88
   Music Bridge trainees 86
   MWB 87
   self reflexivity 87
   sonic effects 86
transformative outside/inside
   perceptions 134–7
*Transformative Power of Performance,
   The* 142
transitional justice 47, 56, 62, 65, 68
Trauerarbeit 6
Tripoli 38, 55, 58
Tripp, C. 54
Troubles, The 23, 122
trust
   absence of 74
   musicking 74–5
   sense of security 74
   uncertainty 74
Truth and Reconciliation Commission,
   South Africa 15

Turino, T. 4
Turner, V. 127

Ubu Roi 24, 123, 139, 141
Ubu the King 24, 123, 137–48, 175
  belliphonic effect 143
  destabilization, dichotomies 140
  ignoble other-self 140
  intrusive noises 141
  liminality 140, 143–4
  noises, troubles 143
  perceptual shifts 142
  sound(ing) war(s) 138–44
  victory speech, Ubu 137, 144
UK, the 123
UK City of Culture 2013 75
United States, the 13, 22, 97, 99–101, 105–6, 118

van Gennep, A. 127
Varoutsos, G. 152
Vasconcellos, J. de 16
Velloso, R.C. 155
Velvet Revolution, the 12
Venezuela 106
Versöhnung 7
Vila do João 156
Vinthagen, S. 48, 51–2
violence 1–2, 6, 11, 14, 18, 20, 31, 35, 38, 56, 59, 66, 89, 93, 98, 101, 103, 109, 122, 124–5, 127, 148–9, 163, 174, 182–3
VJ Day 16
Voegelin, S. 10, 27, 174

voice 3–4, 6, 13, 22–3, 44, 69, 91–2, 99, 101–2, 104, 108–9, 112, 119, 134, 175, 180, 182
  aural characteristics 115
  conflict and conflict resolution 49
  disembodied 145
  dissonant 129–32
  favela 159
  levels of 49, 52
  logocentric 111
  and resistance 49–50
  storytelling 106
  validation 49–50
Voices from the Vault 123
vulnerability 39, 54, 79, 88, 95, 183

wall of shame 152
Wardani, F. 60–1, 63–7
Washington DC 99
West Bank 185
West-Eastern Divan Orchestra 126
With a Hidden Noise (Duchamp) 33
Wlodarski, A.L. 109
word and interpretation 9–11
Writing Culture (Clifford) 3

Yamagishi, T. 75
Yeats, W.B. 139
Yee, S. 24, 123, 144
Yin, R.K. 37
Yousufi, M. 98–104, 108, 112–14, 117–18

Zigon, J. 87–8

www.ingramcontent.com/pod-product-compliance
Lightning Source LLC
Chambersburg PA
CBHW062216300426
44115CB00012BA/2092